Evolving Networking Technologies

Scrivener Publishing
100 Cummings Center, Suite 541J
Beverly, MA 01915-6106

Publishers at Scrivener
Martin Scrivener (martin@scrivenerpublishing.com)
Phillip Carmical (pcarmical@scrivenerpublishing.com)

Evolving Networking Technologies

Developments and Future Directions

Edited by

Kanta Prasad Sharma

GLA University, Mathura, India

Shaurya Gupta

University of Petroleum and Energy Studies, India

Ashish Sharma

GLA University, Mathura, India

Dac-Nhuong Le

Haiphong University, Haiphong, Vietnam

Scrivener Publishing

WILEY

This edition first published 2023 by John Wiley & Sons, Inc., 111 River Street, Hoboken, NJ 07030, USA and Scrivener Publishing LLC, 100 Cummings Center, Suite 541J, Beverly, MA 01915, USA

For more information about Scrivener publications please visit www.scrivenerpublishing.com.

Wiley Global Headquarters
111 River Street, Hoboken, NJ 07030, USA

For details of our global editorial offices, customer services, and more information about Wiley products visit us at www.wiley.com.

Library of Congress Cataloging-in-Publication Data

ISBN 978-1-119-83620-9

Cover image: Pixabay.Com
Cover design by Russell Richardson

Set in size of 11pt and Minion Pro by Manila Typesetting Company, Makati, Philippines

Printed in the USA

10 9 8 7 6 5 4 3 2 1

Dedicated to our friends and family for their constant support during the course of this book

Contents

List of Figures

List of Tables

Foreword

This book discusses some of the critical security challenges facing the ever-evolving networking technologies of today. Chapter 1, 5G Technologies, Architecture and Protocols, presents the main elements in 5G core networks, security in 5G mobile networks, 5G radio access technology, frame structure, network virtualization, and slicing in 5G, which are the key areas of study in 5G technology. Chapter 2, Scope and Challenges of IoT and Blockchain Integration, focuses on the pros and cons of the integration of both the technology and existing platforms that are based on the alliance of IoT and blockchain platforms like Ethereum, Hyperledger, Lisk, and Slock.it, which are also explained along with their full functionality. Chapter 3, Data Communication and Information Exchange in Distributed IoT Environment based on IoT as a paradigm based totally on the internet that contains many interconnected technologies like radio frequency identity and Wi-Fi sensor and actor networks, to exchange data. Chapter 4, Contribution of Cloud-Based Services in Post-Pandemic Technology Sustainability and Challenges, focuses on the contribution of cloud technologies in agriculture, weather forecasting, medical image analysis, security, ICT, and entertainment, along with future application and utilities. This chapter also covers the various applications and tools used by different industrial areas supported by cloud computing. Chapter 5, Network Security in Evolving Networking Technologies: Developments and Future Directions, specifically relates to the network system's security, privacy, integrity and availability of data information in the system. The challenge of network protection persists across all levels of the data network, and the purpose of network security is to protect the secrecy, transparency, integrity, stability, usability and auditability of the network. Chapter 6, The State of CDNs Today and What AI-Assisted CDN Means for the Future, points out the drawbacks of CDN, such as distributed denial of services attacks (DDoS), which are a serious concern for CDN. Chapter 7, Challenges and Opportunities on the problem of the concept of smart cities, which is the need for better technologies to improve the system's data transfer, the research gap there is a problem of trust as massive amounts of the private data of users are at stake, with hackers trying to gain access to it. Chapter 8, Role of IoT in Smart Homes and Offices, discusses the concept of smart offices and homes, which are a part of the smart building environment structure. The role of IoT and cloud computing in establishing communication amongst smart devices is also discussed. It further discusses the components of each technology and the areas of application along with covering future aspects and limitations. Chapter 9, Role of IoT in the Prevention of COVID-19, discusses the challenge currently facing the world, which is how to stop the expansion of the COVID-19 virus as the world is now facing the third wave of the disease, against which the WHO is regularly updating all of us to take precautions against. To prevent the spread of the virus, individuals testing positive for the disease should be placed in isolation. Many countries are currently affected and the entire world is taking precautions against it by using the various methods as per the guidelines issued by the WHO. Chapter 10, Role of Satellites in Agriculture, provides critical reviews of the role of satellites in agriculture and how big data analysis can give amazing results; hence, contributing to the national economy. Further, the advantages and disadvantages along with the challenges that lie ahead are discussed and how the future of these technologies will help the agricultural sector. Chapter 11, Search Engine Evaluation Methodology, discusses how the evaluation concept is a key technology which is used to make continuous and smooth progress in the direction of constructing a better search engine. In the search engine evaluation process, the search engine's performance is measured in respect to its efficiency and effectiveness. Chapter 12, Synthesis and Analysis of Digital IIR Filters for Denoising ECG Signal on FPGA, focuses on the conversion of MATLAB code of different designed IIR digital filters for demising ECG signal into Verilog code using HDL com- mand line interface. Spartan-6 FPGA (XC6SLX75T with 3FGG676 package) is used as a target device. Chapter 13, Neural Networks and Their Applications, discusses how our brain has ten billion cells, which are correspondingly called neurons, that process informa- tion as electric signals.

In closing, we wish to express our sincere thanks to all the authors for their valuable contributions to this volume. Without their cooperation and eagerness to contribute, this project would never have been successfully completed. All the authors have been extremely cooperative and punctual during the submission, editing, and publication process of the book. We express our heartfelt thanks to Martin Scrivener of Scrivener Publishing for his support, encouragement, patience, and cooperation during the entire process of publishing this volume. We would surely be failing in our duty if we do not acknowledge the encouragement, motivation, and assistance that we received from those in India. While it will be impossible for me and my team to mention the name of each of them, the contributions of our reviewers have been invaluable in making this volume as error-free as possible. Last but not the least, we would like to thank all members of our respective families for being the major sources of our motivation, inspiration and strength during the entire time it took to publish this volume.

Kanta Prasad
Sharma Shaurya
Gupta Ashish Sharma
Dac-Nhuong Le
January 2023

Preface

In an age of explosive worldwide growth of electronic data storage and communications, effective protection of information has become a critical requirement. Especially when used in coordination with other tools for information security, cryptography in all of its applications, including data confidentiality, data integrity, and user authentication, is the most powerful tool for protecting information. While the importance of cryptographic technique, i.e., encryption, in protecting sensitive and critical information and resources cannot be overemphasized, an examination of the technical evolution within several industries reveals an approaching precipice of scientific change. The glacially paced, but inevitable convergence of quantum mechanics, nanotechnology, computer science, and applied mathematics, will revolutionize modern technology. The implications of such changes will be far reaching, with one of its greatest impacts affecting information security. More specifically, modern cryptography. With the exponential growth of wireless communications, the internet of things, and cloud computing, and the increasingly dominant roles played by electronic commerce in every major industry, safeguarding the information in storage and while traveling over the communication networks is increasingly becoming the most critical and contentious challenge for technology innovators. The key prerequisite for the sustained development and successful exploitation of information technology and other related industries is the notion of information security and the assurance that operations and information systems are protected by ensuring their availability, integrity, authentication, non-repudiation, information confidentiality and privacy. While it is true that cryptography has failed to provide its users the real security it promised, the reasons for its failure has not much to do with cryptography as a mathematical science. Rather, poor implementation of protocols and algorithms has been the major source of the problem. Cryptography will continue to play a leading role in developing new security solutions that will be in great demand with the increasing bandwidth and data rate of next-generation communication systems and networks. New cryptographic algorithms, protocols and tools must follow up in order to adapt to the new communication and computing technologies. New security mechanisms should be designed to defend against the increasingly complex and sophisticated attacks launched on networks and web-based applications. In addition to classical cryptographic algorithms, approaches like chaos-based cryptography, DNA-based cryptography, and quantum cryptography will increasingly play important roles. However, one must not forget that today's fundamental problems in security are not new. What has changed over the decades is the exponential growth in the number of connected devices, evolution of networks with data communication speed as high as terabits per second, at least in the near field, massive increase in the volume of data communication and availability of high-performance hardware and massively parallel architecture for computing and intelligent software. As the security systems design becomes more and more complex to meet these challenges, a mistake that is committed most often by security specialists is not making a comprehensive analysis of the system to be secured before making a choice about which security mechanism to deploy. On many occasions, the security mechanism chosen turns out to be either incompatible with or inadequate for handling the complexities of the system.

Kanta Prasad
Sharma Shaurya
Gupta Ashish Sharma
Dac-Nhuong Le
January 2023

Acknowledgments

First of all, we would like to thank all our colleagues and friends for sharing our happiness at the start of this project and following up with their encouragement when it seemed too difficult to complete. We are thankful to all the members of Scrivener Publishing, especially Martin Scrivener and Phillip Carmical, for giving us the opportunity to write this book.

We would like to acknowledge and thank the most important people in our lives, our parents and partners, for their support. This book has been a long-cherished dream which would not have become a reality without the support and love of these amazing people, who encouraged us with their time and attention. We are also grateful to our best friends for their blessings and unconditional love, patience, and encouragement.

Kanta Prasad
Sharma Shaurya
Gupta Ashish Sharma
Dac-Nhuong Le

Acronyms

3GPP	3rd Generation Partnership Project
4G	Fourth Generation
4GT	Fourth Generation Techniques
5G	Fifth Generation
6LoWPAN	IPv6 over Low-Power Wireless Personal Area Networks
ABE	Attribute-based Encryption
AES	Advanced Encryption Standard
AHP	Analytical Hierarchy Process
AI	Artificial intelligence
AMPS	Advanced-Mobile Phone System
AMQP	Advanced Message Queuing Protocol
ANN	Artificial Neural Network
API	Application Programming Interface
AR	Augmented Reality
AR/VR	Augmented Reality and Virtual Reality
BDA	Big Data Analytics
BDMA	Beam Division Multiple Access
BLE	Bluetooth Low Energy
BNN	Biological Neural Network
BNS	Bi-Normal Separation
BSS	Business Support System
CBDM	Component-Based Development Model
CCT	Cloud Computing Technology
CCSP	Cloud Computer Service Provider
CDMA	Code-Division Multiple Access
CDMS	Code Division Multiple Access
CDN	Content Delivery Network
CERT	Computer Emergency Response Teams
CIO	Chief Information Officer
CI	Consistency Index
CLEF	Cross Language Evaluation Forum
CMMI	Capability Maturity Model Integration
CN	Core Network

CoAP	Constrained Application Protocol
CPM	Concurrent Process Model
CR	Consistency Ratio
CSA	Cyber Security Agency
CSCM	Cybersecurity Challenges Model
CSS	Cascading Style Sheets
D2D	Device-to-Device
DaaS	Data as a Service
DB	Database
DC2M	DevOps' Culture Challenges Model
DCG	Discounted Cumulative Gain
DCM	Divide and Conquer Model
DCPS	Data-Centric Pub-Sub
DDS	Data-Distribution Service
DDoS	Distributed Denial of Service
DES	Data Encryption Standard
DevOps	Development and Operations
DLRL	Data Local Reconstruction Layer
DMM	Distributed Mobility Management
DNS	Domain Name System
DoS	Denial of Service
DDoS	Distributed Denial of Service
DTLS	Datagram Transport Layer Security
DVB	Digital Video Broadcasting
DX	Digital Transformation
E2E	End-to-End
ECC	Elliptic Curve Cryptography
e-CF	e-Competence Framework
EDGE	Enhanced Data Rate for GSM Evolution
EVM	Ethereum Virtual Machine
FBMC	Filter Bank Multicarrier
FCM	Fuzzy C-Means
FFN	Feedforward Network
FIRE	Forum for Information Retrieval Evaluation
FP	Function Point
FPGA	Field-Programmable Gate Array
FPR	False Positive Rate
FTC	Feature Transition Charts
GSM	Global System for Mobile Communication
GIS	Geographic Information System
GUI	Graphical User Interface

GPU	Graphics Processing Unit
GPS	Global Positioning System
HCI	Human–Computer Interaction
HTTP	Hypertext Transfer Protocol
HR	Human Resource
IaaS	Infrastructure as a Service
IBE	1Identity-Based Encryption
IBFD	In-Band Full-Duplex
ICT	Information and Communications Technology
IDC	International Data Corporation
IDM	Intrusion Detection Manager
IDS	Intrusion Detection System
IMS	IP Multimedia Subsystem
IoE	Internet of Everything
IoT	Internet of Things
I/O	Input/Output
ISO/IEC	International Organization for Standardization/International Electrotechnical Commission
IT	Information Technology
ITU	International Telecommunication Union
KGS	Key Generation Server
KNN	k-Nearest Neighbor
KPI	Key Performance Indicators
KSI	Keyless Signature Infrastructure
LAN	Local Area Network
LED	Light Emitting Diode
LMS	Least Mean Square
LSE	Least Square Estimation
LTE	Long-Term Evolution
M&A	Measurement & Analysis
MD5	Message Digest 5
MES	Manufacturing Execution System
MHT	Merkle Hash Tree
MIH	Media Independent Handover
MIMO	Multi-Input Multi-Output
ML	Machine Learning
MOOC	Massive Open Online Course
MQTT	Message Queuing Telemetry Transport
MSE	Mean Square Error
MVF	Mean Value Function
MVC	Model View Controller

NCS	Non-Cognitive Skills
NCSF	Non-Cognitive Skills Framework
NFC	Near-Field Communication
NFV	Network Feature Virtualization
NFV-MANO	Network Feature Virtualization Management and Orchestration
NGMN	Next Generation Mobile Network
NMT	Nordic Mobile Telephone
NPS	Net Promoter Scores
NSSI	Network Slice Subnet Instance
OFDM	Orthogonal Frequency-Division Multiplexing
OSI	Open Systems Interconnection
OSS	Open-Source Software
OS	Operating System
OTT	Over-the-Top
OWASP	Open Web Application Security Project
P2P	Peer-to-Peer
PaaS	Platform as a Service
PIC	Programmable Microcontroller
PIO	Population, Intervention, and Outcome
PPC	Pay-per-Click
PU	Perceived Usefulness
PUF	Physical Unclonable Function
PV	Planned Value
QA	Quality Assurance
QoS	Quality of Service
RAD	Rapid Application Development Model
RAT	Radio Access Technology
RAN	Radio Access Network
RC4	Rivest Cipher 4
RF	Radio Frequency
RFC	Request for Change
RFID	Radio Frequency Identification
RM	Risk Management
RPM	Rapid Prototyping Model
RNG	Random Number Generator
RSA	Rivest-Shamir-Adleman
RTT	Round-Trip Time
RT-PCR	Real-Time Polymerase Chain Reaction
SaaS	Software as a Service
SBA	Service-Based Architecture
SBR	Security Bug Reports

SDN	Software-Defined Networking
SEO	Search Engine Optimization
SHA	Secure Hash Algorithm
SOA	Service-Oriented Architecture
SPSS	Statistical Package for Social Sciences
SQL	Structured Query Language
SME	Small and Medium Enterprises
SMS	Short Message Service
SV	Schedule Variance
SVM	Support Vector Machine
SLR	Systematic Literature Review
SSE	Sum of Squares Error
SSL	Safe Socket Layer
SWIPT	Simultaneous Cellular Data and Power Transfer
TACS	Total Access Communication System
TCP/IP	Transmission Control Protocol/Internet Protocol
TCO	Total Cost of Ownership
TLS	Transport Layer Security
TSP	Traveling Salesman Problem
UDP	User Datagram Protocol
UMTS	Universal Mobile Telecommunications System
URL	Uniform Resource Locator
VR	Virtual Reality
XML	Extensible Markup Language
XMPP	Extensible Messaging and Presence Protocol
XP	Extreme Programming
XSS	Cross-Site Scripting
ZKP	Zero Knowledge Proof
W3C	World Wide Web
WBS	Work Breakdown Structure
WiMAX	Worldwide Interoperability for Microwave Access
WSM	Win-Win Spiral Model
WAN	Wide Area Network
WSN	Wireless Sensor Network
WWW	World Wide Web
WWWW	World Wide Wireless Web

1

5G Technologies, Architecture and Protocols

Shweta Bondre[1], Ashish Sharma[1], Vipin Bondre[2]

[1] GH Raisoni College of Engineering, Nagpur, India
[2] Yeshwantrao Chavan College of Engineering, Nagpur, India
Email: shweta.kharat@raisoni.net,ashishk.sharma@raisoni.net, vdbondre@ycce.edu

Abstract

The term "5G" refers to the fifth generation of mobile technology. With each passing day, this field of telecommunications has seen a number of changes, from the first generation to 2.5G, and from 3G to 5G, with better efficiency. This growing mobile technology revolution is transforming our daily lives, including how we chat, work, learn and so on. The fifth generation network provides affordable mobile internet access at very high speed. The study aims to shed light on fifth generation technology network architecture and protocols. The development of the world wide wireless web (WWWWW), wireless networks, actual wireless and complex ad hoc world in the fifth century are all being studied. The 5G technology is based on voice-over IP technologies that provide users with a high level of call volume and data transmission. The ability to connect to and switch between various wireless networks at the same time is one of the main characteristics of the 5G mobile network. The main elements in 5G core networks, security in 5G mobile networks, 5G radio access technology, frame structure, network virtualization, and slicing in 5G are the key areas of study in 5G technology.

Keywords: 5G, wireless technology, network architecture and protocols

1.1 Evolution of Wireless Technologies

Wireless-based networks will continue to develop in a number of respects now and in the coming future to address new demands and threats. New technology elements, such as high-speed packet communication and long-term growth, will be launched as part of the development of existing cellular-based networks. Mobile wireless networking has progressed from analog voice calls to modern emerging technologies capable of providing high-quality mobile broadband networks with end-user data rates of several megabits per second over wide regions. The immense advances in the potentiality of mobile communication networks, along with the introduction of advanced models of mobile devices, such as smart phones and tablets, have resulted in a proliferation of emerging mobile networking technologies and an exponential increase in network traffic. Our vision for the future is a connected society with free access to information and data that is accessible to all at all times.

New technology elements must be investigated for the evolution of already available wireless-based systems in order to achieve this vision. Existing wireless systems, such as Wi-Fi, HSPA and 3GPP-3rd Generation Partnership Project, LTE technology will incorporate emerging technology elements to better meet future needs. In comparison to current 4G LTE networks, the ultimate aim of the upcoming 5G wireless networking is to have comparatively high download rates, very low latency, significant increases in base-station reliability and significant improvements in expected quality of service (QoS) for consumers. Broadband data consumption has grown at a rapid rate because of recent technology and networking in the context of internet of things (IoT) programs, smart mobiles, autonomous cars, smart home connectivity and augmented reality devices, etc.; therefore, in order to support the most recent applications, the system's bandwidth must be significantly expanded. This advancement could be made possible by the use of a modern spectrum and higher data volumes.

The history of wireless network technology is given below [1] and is depicted in Figure 1.1.

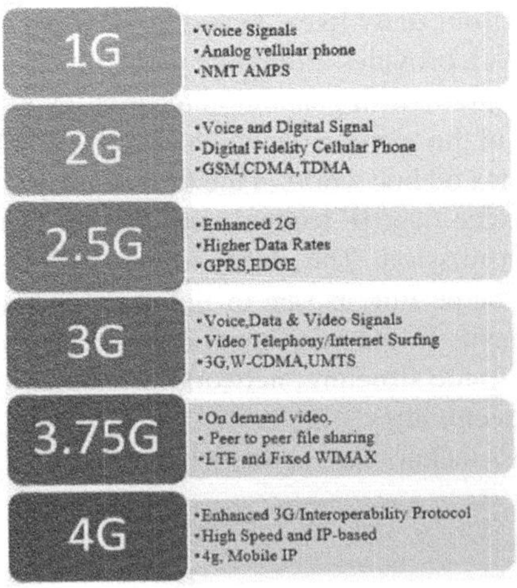

Figure 1.1: Evolution of wireless technologies.

In terms of range, spectral quality, mobility, and data rate, it reflects the progression of wireless technology generations. It also shows that circuit switching is used in 1G and 2G technologies, and both circuit and packet switching is used in 2.5 Generation and 3 Generation, and packet switching is used in the subsequent generations from 3.5 Generation to now, i.e., 5 Generation. Following is an overview of the emerging wireless technologies:

- **1G**: In the early 1980s, the first generation was announced. It has a 2.4 kbps maximum data limit. Total Access Communication System (TACS), Advanced Mobile Phone System (AMPS) and Nordic Mobile Telephone (NMT) were among the top subscribers. It has a number of drawbacks, including reckless handoff, insufficient capacity, lack of security and poor voice associations as voice calls are kept and played in radio towers, increasing the risk of uninvited eavesdropping by 3rd parties [2].
- **2G**: The second generation was launched in the late 1990s. In second-generation mobile phones, digital technology is used. The first second-generation system, Global Systems for Mobile Communications (GSM), was used for voice communication and had a data rate of up to 64 kbps. Because radio signals are low in power, 2G mobile handset batteries last longer. E-mail and Short Message Service (SMS) are among the services it offers. IS-95, Code Division Multiple Access (CDMS) and GSM were all important systems [2,3].
- **2.5G**: It usually utilizes a second-generation cellular infrastructure that provides General Packet Radio Services (GPRS) and other technologies not available in 2G or 1G networks. A 2.5G method, in contrast to a 2G method, uses packet switching in addition to circuit switching. It supports data rates of up to 144 kbps. The key 2.5G technologies were Enhanced Data-Rate for GSM Evolution (EDGE), GPRS, and Code-Division Multiple Access (CDMA) 2000 [2,3].
- **3G**: In late 2000, the third generation was introduced. It transmits data at a speed of up to 2 Mbps. High-speed broadband connectivity is paired with internet protocol-based applications in 3G technology (IP). In addition to transmission rate, an additional upgrade was made to retain QoS. The 3G technology was distinguished by additional features such as worldwide roaming and improved speech quality. The only downside of to 3G mobile sets is that they use considerably more electricity than 2G units. Furthermore, 3G network services are more costly than 2G network plans [2,3]. Since 3G includes the introduction and use of Wideband Code-Division Multiple Access (WCDMA), Code-Division Multiple Access (CDMA) 2000 and Universal Mobile Telecommunications-Systems (UMTS) technologies, emerging technologies such as Evolution-Data Optimized (EVDO) and High Speed Uplink/Downlink Packet Access (HSDPA/HSUPA) have produced a 3.5G wireless generation which is an intermediate between 3G and 4G.
- **3.75G**: Fixed Worldwide Interoperability for Microwave Access (WiMax) and Long-Term Evolution (LTE) are the promise of mobile communication networks. Fixed WiMAX and LTE will supplement network bandwidth by allowing a large number of users to connect to high-speed networks like peer-to-peer data sharing, on-demand streaming, and composite Web services. An additional spectrum is now available, allowing operators to operate their networks more compliantly and with greater coverage and capacity at a lower cost [2,4].
- **4G**: The 4G standard is considered a descendant of the 2G and 3G standards. Worldwide Interoperability for Microwave Access (WiMAX), in collaboration with Mo-

bile WiMAX, the 3rd Generation Partnership Project (3GPP), is now standardizing Long-Term Evolution (LTE), advanced as a long-term 4th Generation standard. A 4G framework enhances existing connectivity networks by providing a comprehensive and reliable IP-based solution. Data, multimedia and voice can be delivered to customers at all times and in all places, and at even higher data speeds than previous generations. Multimedia messaging service (MMS), high-definition TV content, digital video broadcasting (DVB), video chat and mobile TV are some of the applications that are being improved to use a 4G network [1,4,5].

- **5G**: With the exponential growth in customer demand, 4G can be rapidly replaced by 5G by using advanced access technologies such as non- and quasi-orthogonal or filter bank multicarrier (FBMC) multiple access and beam division multiple access (BDMA). To understand the idea behind the BDMA technique, consider the situation of a base station interacting with mobile stations in which every cellular mobile station is given an orthogonal beam, and the BDMA method splits the antenna beam according to the positions of the mobile stations to provide various accesses to the mobile stations, thus increasing power [6]. Based on recent trends, it is widely believed that 5G cellular systems must overcome six issues that 4G cellular networks cannot successfully address: greater bandwidth, higher data rate, lower end-to-end latency, higher data rate, lower cost, large device connectivity and consistent quality of experience provisioning [7,8]. In mobile technology, a 5G network is thought to be the peak of cellular networking. Cell phones are useful for a variety of activities in addition to messaging. Both previous wireless devices have made it easier to communicate and share data, but 5G adds a different dimension to the experience, turning it into a true smartphone experience.

Table 1.1: Evolution of wireless technology.

Generation	Time Period	Technology	Speed	Switching	Features
1G	1970-1980	TACS,MPS,NMT	14.4 Kbps	Circuit	Wireless phones are only used for speech during 1G.
2G	1990-2000	CDMA, TDMA	9.6/ 14.4 Kbps	Circuit	Multiplexing allows multiple devices to use a single channel in 2G functionality. Cellular phones were used for data as well as voice during the 2G era.
2.5G	2001-2004	GPRS	171.2 Kbps 20-40 Kbps	Circuit/Packet	2.5 gallons Data is becoming more important as the internet becomes more popular. 2.5 gallons Broadcasting and multimedia networks are becoming increasingly popular. Web browsing became available on phones.
3G	2004-2005	WCDMA, UMTS, EDGE, CDMA 2000	3.1 Mbps 500-700 Kbps	Circuit/Packet	Multimedia services, as well as streaming, are more popular on 3G. 3G enables universal access and portability across a variety of device types.
3.5G	2006-2010	HSPA	14.4 Mbps 1-3 Mbps	Packet	Higher throughput and speeds are supported by 3.5G, allowing consumers to meet their increased data demands.
4G	NOW	WiMax,,LTE, Wi-Fi	100-300 Mbps. 3-5 Mbps 100 Mbps (Wi-Fi)	Packet	4G speeds are being increased even more to keep up with data access demand from various services. 4G now supports high-definition streaming. New HD-capable phones are on the market. The portability of 4G is even greater.

1.2 5G Cellular Network Architecture

Figure 1.2 shows a schematic diagram of the broadband and mobile interoperability of the device architecture for 5G mobile systems. In the infrastructure, there is a user terminal (that plays a significant role in modern design) as well as a variety of independent, autonomous radio connectivity technologies. Inside each terminal, any of the radio access technologies is viewed as an IP link to the outside internet world. Each radio access technology (RAT) should, however, have its own radio interface in the mobile station. For example, if we wish to bind to four different RATs, the mobile terminal would require four different access-specific interfaces, all of which must be active at the same time for this architecture to function.

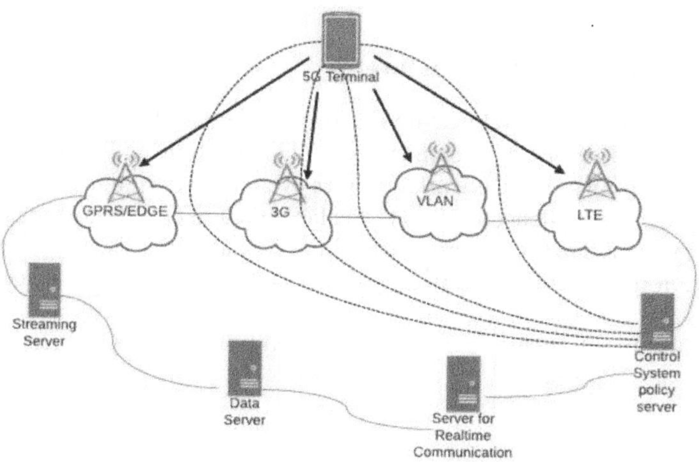

Figure 1.2: Functional architecture for 5G mobile networks.

The first two OSI layers (data link layer and physical layer) define the radio access technologies that enable users to connect to the internet by QoS support, which is based on the access technology. The network layer sits on top of the OSI-1 and OSI-2 layers in today's networking environment, and it is IP (Internet Protocol), either IPv4 or IPv6, regardless of the equipment used for radio connectivity. IP's goal is to ensure that sufficient control information in the IP header is available to ensure proper routing of IP packets belonging to specific device links/sessions between client applications and servers located anywhere on the internet. Packet routing should be done in compliance with the user's defined policies. On the internet, sockets are used to provide connections between clients and servers. Internet sockets are the endpoints for data transfer flows. Each web socket is a one-of-a-kind combination of a local network communications port and IP address, a target communications port and IP address, and a transport protocol. End-to-end communication using the internet protocol is required between the client and server in order to lift the necessary internet socket, which is uniquely decided by the client and server's application. This means that the destination IP address and local IP address should be set and unchanged in the case of interoperability between heterogeneous networks and vertical handover among radio technologies. When these two conditions are set, when a smartphone user is present on at least one end of the network, the internet connection should have end-to-end handover clarity. An IP interface is provided for each radio access technique that the user has access to in order to connect to the relevant radio access. Each IP interface in the terminal has its own IP address, netmask, and network parameters for IP packet routing. Changing the access technology means changing the local IP address in a regular inter-system handover. The socket is then changed by changing some of its parameters, resulting in the socket being closed and a new one being opened. This means that the connection will be terminated and a new one will be created. To address this shortcoming, a new layer will be responsible for the abstraction stages of network access technology to upper layers of the protocol stack. In this work, the authors implement a control mechanism in the functional design of the networks, which operates in full synchronization with the user terminal and offers network abstraction functionality and packet routing based on the most appropriate radio access technology, to allow the functions of implemented clarity and control or direct routing of packets via the most appropriate radio access technology [9].

1.2.1 5G E2E Network Architecture

The 5G end-to-end network architecture is depicted in Figure 1.3 below, which outlines the 5G E2E network. However, by switching from "4 Generation" to "5 Generation", the E2E structural design of the "5G" network becomes even more essential since the base station is no longer the key bottleneck [10,11]. Scalable data exposure governance and access management systems are used to provide facilities for data analysis, collection, distribution and abstraction on a shared network where data can be accessed by device entities at all levels.

The architecture of Huawei's end-to-end network built for 5G is represented in Figure 1.3.

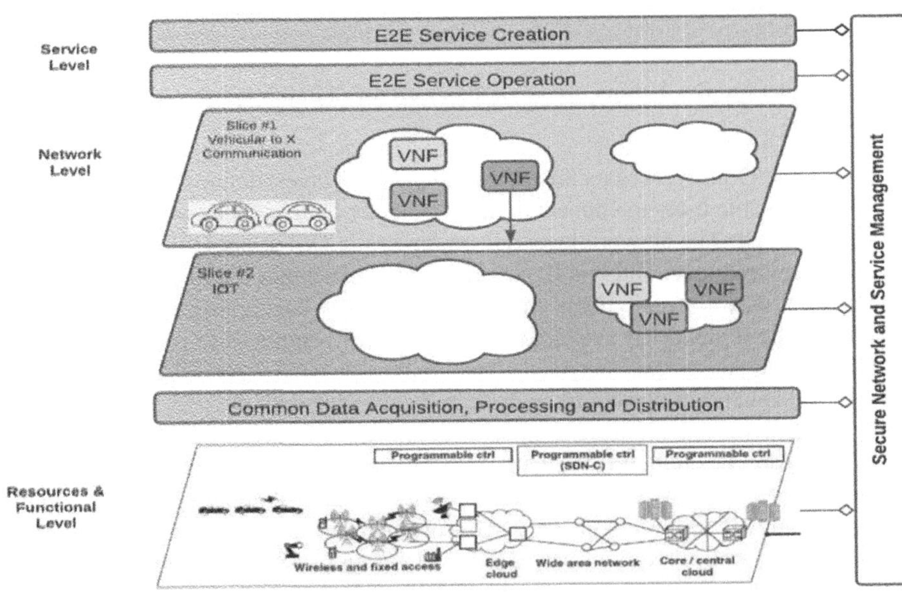

Figure 1.3: Architecture of end-to-end network in 5G [11].

1.2.2 Network Slicing Architecture

Network slicing helps you run several dedicated networks on a single platform, which is a very powerful process. Network slicing is an example of the idea of easily and cost-effectively operating multiple logical networks as essentially autonomous business processes on a single physical infrastructure. It is a 5G cutting-edge technology that can build several types of virtual networks, tailored to meet different specifications for various use cases. Network slicing architecture provides a number of independent service-level arrangements to satisfy the requirements. A network slice is divided into two types: CN network slice subnet instance (CN NSSI) and RAN network slice subnet instance (NSSI). Network slice instance and Resource layers are shown in Figure 1.4.

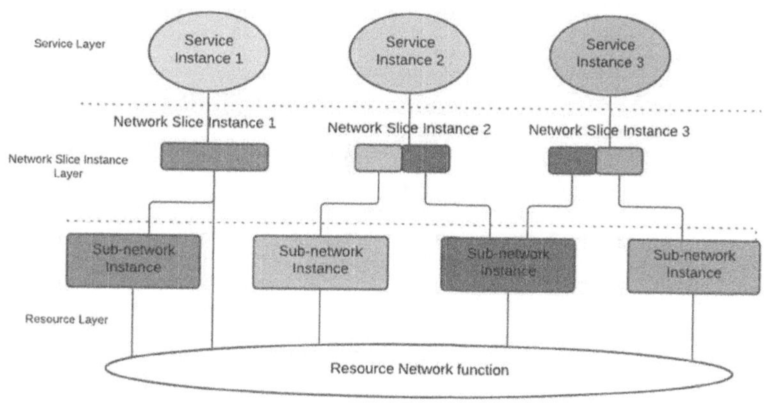

Figure 1.4: Component of network slicing [12].

The end user or enterprise services provided by the network are represented by the Service instance layer. The network characteristics specified by the service instance are provided by the Network slice instance. Multiple service instances can share a single network slice instance. A set of network functions that operate on the computational, physical or virtual Resource layer is referred to as a sub-network instance [12].

Network slicing includes slicing in radio access network (RAN) and in core network (CN). Software-defined network (SDN) and network feature virtualization (NFV) are technical enablers for network slicing in CN because NFV and SDN virtualize and manage network components and functions, allowing for simple customization and reuse of certain elements and functions in each slice to satisfy service requirements. Slicing might be based on physical or logical radio resources abstracted from physical resources on the RAN side. Slices of the network would need to be of a variety of shapes and sizes. This necessarily requires a high degree of flexibility [13]. According to the commercial purpose of a given slice for a specific industry or the user/machine experience that they are designed to serve, network slicing can also be divided into vertical slicing and horizontal slicing.

1.2.3 NFV Management and Orchestration

Virtual network functions (VNFs) can be reassigned and deployed to share the infrastructure's virtual and physical resources, ensuring scalability and performance requirements. Telecom Service Providers (TSPs) will easily launch new and elastic services as a part of this [14,15]. Services, NFVI, and NFV management and orchestration (NFV-MANO) are the three key components of the NFV architecture, as seen in Figure 1.5.

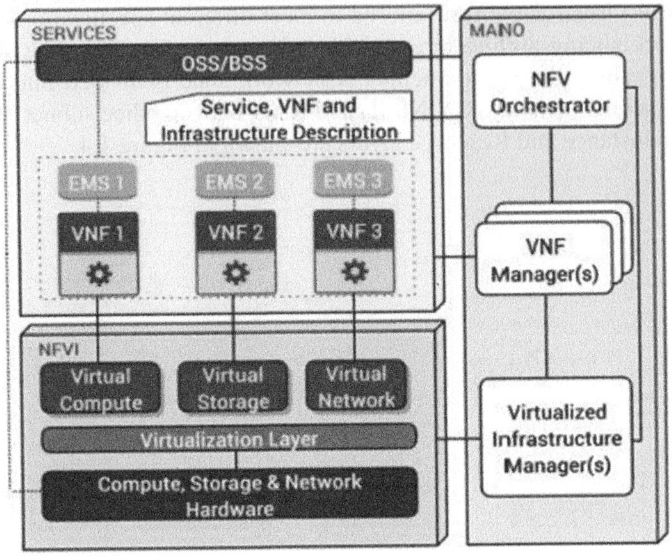

Figure 1.5: ETSI-NFV architecture [16].

Virtual network functions (VNFs): A collection of VNFs which can be developed in one or more virtual machines is referred to as a utility. In some cases, VNFs can be executed in virtual machines built into operating environments or directly on hardware; in these cases, native hypervisors or virtual machine monitors are used to manage them. The OSS, as well as the business support system (BSS), is a general management system that assists operators in deploying and managing a variety of E2E telecommunications networks (e.g., problem troubleshooting, renewals, billing, ordering, etc.). The attention of the NFV requirements is on integration with current OSS/BSS systems [17].

- NFVI: Both hardware and software capabilities that make up the NFV ecosystem are covered by NFV infrastructure. NFVI covers network access to locations such as data centers and public or private hybrid clouds. The virtualization layer, which sits just above the hardware and abstracts the physical resources, provides storage, encoding, and networking for VNFs. A current virtualization layer, such as a hypervisor, can be used in an NFV implementation with simple features that simply removes hardware devices and transfers them to the VNFs.

- NFV-MANO: The orchestrator, virtualized infrastructure managers, and VNF managers make up NFV management and orchestration. These blocks offer the features needed for management tasks such as provisioning and initialization of VNFs. The orchestration and life-cycle management of physical and software devices that allow infrastructure virtualization, as well as the life-cycle management of VNFs, are the responsibilities of NFV-MANO. It also includes databases for storing data and data models that describe function, facility, and resource implementation and life-cycle properties. The specification also defines interfaces for coordination between the different components of the NFV-MANO, as well as integration by traditional network management systems (e.g., OSS and BSS) to make it easier to execute all NFs and functions on legacy equipment [17,19].

1.2.4 NGMN Envisioned 5G Design

The NGMN Alliance (Next Generation Mobile Networks Alliance) envisions an architecture based on the design principles that takes advantage of the hierarchical separation of software and hardware, along with the programmability provided by NFV and SDN. The 5G-architecture is a native "SDN/NFV" architecture that covers all features of the 5G framework, including equipment, (mobile/fixed) networks, network features, value supporting functionality, and all management functions. This architecture is illustrated in Figure 1.6.

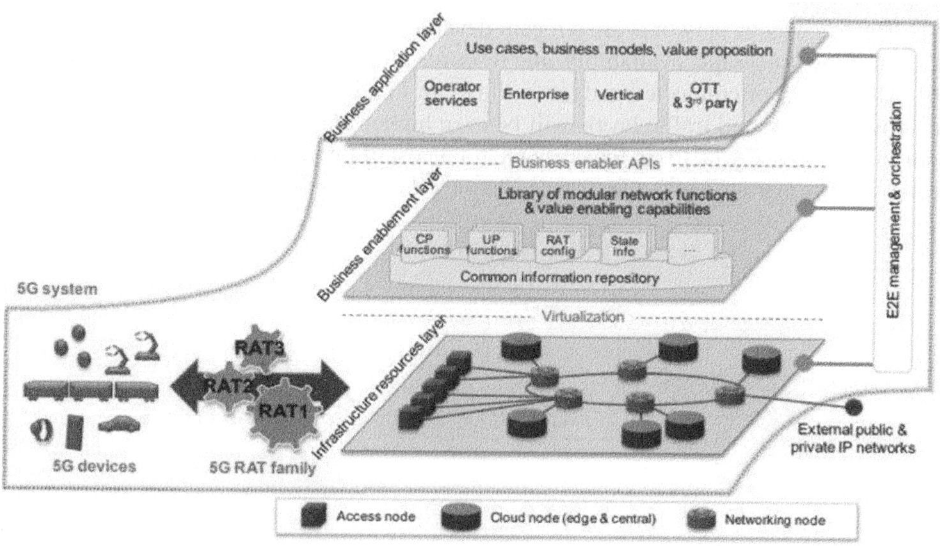

Figure 1.6: 5G architecture [20,21].

Access nodes, wearables, cloud nodes, CPEs, 5G devices, phones, networking nodes, and machine-type modules and associated links make up the infrastructure resource layer of a fixed-mobile converged network. As a result, 5G devices are included in the configurable infrastructure resource. The resources are visible to higher levels as well as the end-to-end management and orchestration organization through related APIs. Monitoring output and status, as well as settings, are all required features of such an API.

The business enablement layer is a modular architecture building block library that contains all network functions available, including functions realized by software modules that can be downloaded from a registry and implemented at the desired location, as well as a collection of configuration parameters for particular network components, such as radio access.

The business application layer has specific software and facilities used by the provider verticals, enterprise or third parties on the 5G network. You can build dedicated network slices for an application or map an application to existing network slices using the end-to-end control and orchestration entity's interface.

1.3 5G Energy Efficiency

In comparison to current 4G systems, the current 5G structure suggests that energy consumption can be reduced by ten percent to extend battery life, in addition to lowering the power consumption desired for wireless base station antenna and client devices [22]. Energy consumption is now a key component in the design of communication networks, and networks are being developed based on this factor [23]. Due to technical advances, internet traffic nowadays is growing every day, and as a result, the "round-trip time" latency of the data packets is increased in the network [24], which is becoming a more important problem in relation to energy prices on 5G networks. In the telecommunications industry, cell systems are the primary cause of increased energy consumption [25]. Rapid electricity usage is a significant barrier to reaching green environmental goals and lowering device costs. Heterogeneous networks are a recent development that is growing in popularity as a way to improve coverage, availability, and resource efficiency in the upcoming 5G network [26]. The variety of connected gadgets will be ten to a hundred times greater than it is in the network today, and traffic levels will be even higher than they are now. The network's electricity use is a key factor in lowering the total cost of ownership (TCO), which includes the network's environmental effects (see Figure 1.7).

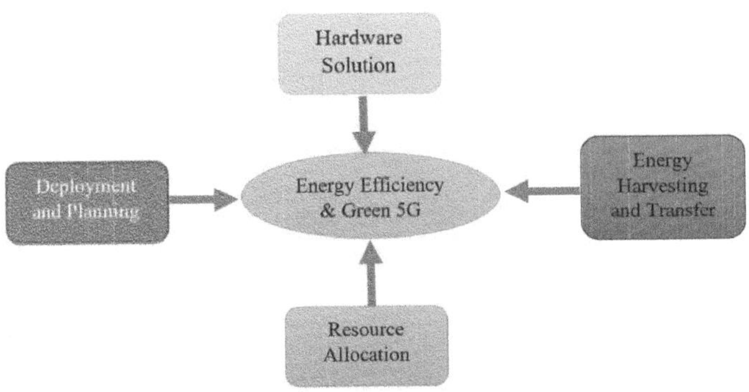

Figure 1.7: 5G energy efficiency.

There are some factors that contribute to the 5G network becoming an energy-efficient system in the coming future [27,28], a few of which are given below.

1.3.1 Full Duplex

The 4G mobile network heavily relies on orthogonal frequency-division multiplexing (OFDM) as a technology for physical layer [30] as a result of its higher band consumption and heavy fading potential [29]. The physical layer of 5G mobile connectivity, on the other hand, is said to have higher demands for scalability, stability, durability, bandwidth performance and robustness [31]. Specialized signal processing electronics now allow full-duplex network communications at the same frequency [32]. For the following purposes, full-duplex wireless networking enables transmitting and receiving on the equal frequency spectrum at the exact time of "radio," as it is one of the candidate techniques outside of 5G and wireless networks. The advantages include the potential for increased bandwidth and improved spectrum performance. Nonetheless, one of the main problems of absolute duplex technology is preventing strong self-interference [33]. The author of [34] introduced a 5G and higher technology that incorporates huge multi-input multi-output (mMIMO) and in-band full-duplex (IBFD) technology. With the same time-frequency resources, IBFD mMIMO can accommodate a significant number of uplink and downlink consumers, massively increasing system energy [31]. Because of the huge rise in the number of antennas, IBFD mMIMO will decrease the complexity of the base station design [35]. The authors suggested the IBFD mMIMO architecture as a central breakthrough to promote easy progress towards future 5G and higher networks [29,36] because of these favorable circumstances.

1.3.2 High Network Data Rate

The 5G networks provide more information and have lower packet latency [37]. They will leave the network linking among the base station and the client disabled for a long period of time. Longer sleeping modes are possible because of these idle hours. Since extraordinarily fast data rates can be given for cellular smartphones, "millimeter-wave telecommunication" is a promising technique for potential 5G wireless networks [38]. The writers proposed the use of millimeter wave telecommunication technologies for D2D communication over the network in [39-41]. The authors in [42] propose a space division multiplexing technology that would maximize bandwidth capability while lowering energy consumption for higher data speeds.

1.3.3 Dense Small Cell Deployment

To attain uninterrupted coverage in city areas and shape a 5G ultradense wireless network, a substantial number of small cells must be deployed [60]. Small cells come in a number of sizes based on how they are classified:

1. Femtocells
2. Picocells
3. Microcells

Small cells can be linked to the network's core by a remote radio header or a central base station, which can be wireless or wired. This reduces the distance among the user and the base station, lowering the transmission power required to overcome the no-path, particularly in an indoor environment, boosting uplink and downlink communications' energy efficiency. Dense small cell deployment is needed to increase signal power and offload macrocells [43]. Furthermore, unregulated small cell deployment may result in uncontrolled cell shapes, leaving network operators with little control over small cell placement [44]. Through using small cells, both indoors and outdoors, we will provide a cost-effective, simple method to the network capability issue created by the huge rise in cellular mobile traffic. The introduction of small cells with a limited radius is expected to increase spectrum and network performance [45].

1.3.4 Massive MIMO Antennas

Cellular network energy consumption is becoming particularly critical for mobile network providers [46] because it has an important economic and environmental effect on next-generation broadband networks, such as the 5G network [47]. The 5G network will almost certainly be hundred a times quicker than the present 4G network. Trying to meet this willing target using ideal models and processes from existing programs is impractical and would almost inevitably result in an energy crisis with serious environmental and financial implications. Huge MIMO will improve the performance of wireless transmission systems' bandwidth by more than 10 times [48]. Any automated and hybrid precoding plans have been highlighted in [49] to increase the energy efficiency of MIMO communication. These innovations are expected to satisfy the demands of rising data rates by using the space domain more efficiently. Massive MIMO is a cutting-edge advancement over existing MIMO technologies. The basic aim of cutting-edge technologies is to distinguish the benefits of MIMO for a broader variety of applications by increasing throughput, spectrum efficacy, and energy consumption while reducing the multifaceted nature into a precoder/identifier [50]. Huge "multi-input multi-output (MIMO)" is a new technique that extends the MIMO technique [51,52]. The simultaneous wireless information and power transfer (SWIPT) technology developed by the authors [53] aims to increase energy efficiency.

1.4 Security in 5G

The new 5G wireless networks will have better coverage, significantly increased quality of service (QoS), low latency, and extremely fast data rates. 5G will also offer ultra-reliable and inexpensive wireless connectivity to cellular handheld devices and cyber-physical networks, as well as a wide variety of modern devices connecting to the internet of things (IoT), universal M2M, and ultra-reliable and affordable broadband access to cellular handheld devices and cyber-physical systems [54]. The protection of 5G is becoming more important because of the probable position of 5G and its effect on our lives. As a result, significant steps are expected to guarantee the reliability of the 5G network infrastructure, its customers, and the 5G network itself [55]. On the other hand, 5G would require the advancement of all network elements, such as central and control networks, as well as all protocol levels from radio to applications [56]. As a consequence, protection could be compromised anywhere (see Table 1.2).

Table 1.2: 1G to 4G security mechanism.

	1G	2G	3G	4G
Security Mechanism	There is no explicit security or privacy mechanism in place.	Protection based on anonymity, authentication, and encryption	2G security was implemented, as well as secure network access, authentication and key agreement, and two-way authentication.	New encryption and trusted mechanisms were introduced, as well as security for encryption keys, integrity protection and non-3G partnership project access security.
Security Challenges	Call interception, eavesdropping an d a lack of privacy protection	Spamming, a fake base station, radio link security, and one-way authentication	IP traffic protection, encryption key security, and roaming security all have flaws.	Security problems triggered by increased IP traffic include DoS threats, data-integrity, base-transceiver station security, and eaves-dropping on long-term keys. Security of new services & computers, such as the huge IoT that 5G offers, is not suitable.

According to the International Telecommunication Union (ITU) [57], security features are logically divided into separate architectural components by a security architecture. This enables a systematic procedure for E2E service security, which aids in the planning of current network security assessments and the implementation of new security results. The security design for 5G has been described as follows:

i) Security of network access: A collection of security parameters that allow user tools to securely authenticate and access network devices. System protection necessitates the monitoring of "3GPP" and "non-3GPP" contact networks, as well as the transition of security contexts from the SN to the user equipment.

ii) Network domain security: It has a number of security features that allow network nodes to share signaling and user-level data in a safe manner.

iii) User domain security: Allows users to view user equipment in a safe way, which protects them.

iv) Application domain security: Applications (from both the customer and vendor domains) will easily exchange messages thanks to encryption software.

v) Service-based architecture (SBA) domain-security: The security functions include network element registration, discovery, and authorization, as well as the security of service-based interfaces.

vi) Security visibility and configurability: This involves letting the user know if the protection feature is turned on.

1.5 5G Applications

1.5.1 Rapid Data Transmission

The 5G network is defined by its high broadband speeds and smart networks. A 4G feature film takes about eight minutes to download; with 5G, in less than five seconds, people would be able to do so. Multimedia television, high-resolution and 3D content, robotics, social networking sites, augmented reality, advanced manufacturing, driverless vehicles and other technologies can all benefit from increased network speed. Not all data must be transferred simultaneously across the billions of computers that will be connected. Some systems require real-time communication, while others can be shared during off-peak hours. In the 5G world, networks must be able to handle data traffic in real time and make split-second decisions [60]. Scientists predict that in this connected world, a much higher proportion of digital information of 35% will be used compared to the previous 5% [61]. Wireless technology research is now looking into a number of possibilities for a future wireless network. The upcoming 5G system should prioritize high-speed connectivity and low-latency specifications. You can stream 8K or ultra-3D videos in a fraction of the time thanks to 5G, which has 40-times higher quality than 4G [62,63].

1.5.2 5G Flexibility for Smart Mobility

In the context of 5G heterogeneous wireless networks, particularly vehicle networks, we intend to combine the MIH paradigm with DMM approaches in the future. Additional parameters, such as vehicle speed, network size, latency and the likelihood of failure to produce, should be cautiously considered in this kind of network defined by a great agility environment [64]. The 5G transportation infrastructure includes everything from traditional road/route planning to new autonomous driving technologies (connected vehicles) and smart-transport sharing. Road management, incident avoidance, secure navigation, fuel conservation, emission reduction and price reduction are all advantages of smart mobility [65,66].

1.5.3 5G in Smart Cities

In the near future, 5G technology will connect the world, from the largest megacities to the tiniest internet of things, in an ever online fashion. Smart homes, smart cities and the IoT will all be merged into one big cohesive infrastructure as a result of this linked hierarchy [67]. Researchers have looked at smart cities and self-organizing networking policies for 5G wireless networks. Smart cities rely on 5G to allow widespread M2M communications, but the network is completely oblivious to the data that is flowing through it. Furthermore, 5G is expected to bring together a variety of connectivity networks, significantly enhancing the reliability of the communication network and making knowledge exchanges between heterogeneous systems and services faster [68]. In smart cities, the tactile internet will provide a platform for measuring, monitoring, recording, and scaling smart devices in physical or virtual reality [69]. The tactile internet's main characteristics are ultra-low latency, reliability, and connection quality, which make it more advanced in 5G [70].

1.5.4 5G Augmented Reality

In current years, augmented and virtual realities have begun to benefit from video streaming technologies and cellular networks' high-speed capacities. Constraints like bandwidth and latency, on the other hand, prohibit us from achieving "high-fidelity telepresence" and advanced virtual and augmented reality technologies. Fortunately, all developers and engineers are mindful of these challenges, and 5G networks have been designed to support us in moving to the next generation of user interfaces [71]. The internet of things and wireless internet are the two major industry drivers for potential cellular networking growth, which will offer a broad variety of possibilities for 5G. In the 5G age, there will be a wide variety of technologies, including augmented and virtual reality, wireless networking, e-Health networks, and car driving, among others [72,73]. Despite the network demands of emerging technology audiences like augmented and virtual reality, there is too much excitement and anticipation for the launch of "5G" network technologies. Smartphone augmented reality and virtual reality (AR/VR) is predicted to be among the first wave of 5G applications. The global AR market is predicted to reach $114 billion by 2021, according to the ABI Report, while the global virtual reality industry is about to reach $65 billion at the same time [71]. Users may provide immersive interactions with both AR and VR; however, they need connectivity that can guarantee high-quality 360° video, low-latency two-way communications, and effective localization. Because of 5G, those drills will be possible on consumer electronics, remotes, and handheld devices, resulting in a multitude of novel instructional scenarios [74-80]. The internet of things and device-to-device networking are two examples of conventional and digital technologies that 5G networks aim to serve.

1.6 Conclusion

In this study we analyzed numerous aspects of the future "5G" network and addressed several architectures which are based on it. 5G networks can link everything together, from a person to the internet, from a basic sensor device to a sophisticated self-driving system, from embedded sensors in all types of hardware to autonomous vehicles, from an airplane to smart businesses and cities. In comparison to today's network, the 5G network has significantly higher network capacity, lower latency, and significantly higher bandwidth. It not only has the potential to change people's lives, but it also has the potential to save them by improving emergency care and reducing traffic accidents. Prior to the commercialization of 5G technology, it is critical to continue to improve network capability. In this chapter we covered the evolution of 5G wireless technology, architectures, energy efficiency in 5G, 5G security and extensive uses of 5G technology in our daily life.

References

1. Rappaport, T. S. (1996). *Wireless Communications: Principles and Practice* (Vol. 2). New Jersey: prentice hall PTR.
2. Santhi, K. R., Srivastava, V. K., SenthilKumaran, G., & Butare, A. (2003, October). Goals of true broad band's wireless next wave (4G-5G). In *2003 IEEE 58th Vehicular Technology Conference. VTC 2003-Fall (IEEE Cat. No. 03CH37484)* (Vol. 4, pp. 2317-2321). IEEE. DOI: 10.1109/vetecf.2003.1285943
3. Halonen, T., Romero, J., & Melero, J. (Eds.). (2004). *GSM, GPRS and EDGE performance: evolution towards 3G/UMTS*. John Wiley & Sons. DOI: 10.1002/0470866969
4. Andrews, J. G., Ghosh, A., & Muhamed, R. (2007). *Fundamentals of WiMAX: understanding broadband wireless networking*. Pearson Education.
5. Sesia, S., Toufik, I., & Baker, M. (2011). *LTE-the UMTS long term evolution: from theory to practice*. John Wiley & Sons.
6. Wang, C. X., Haider, F., Gao, X., You, X. H., Yang, Y., Yuan, D., ... & Hepsaydir, E. (2014). Cellular architecture and key technologies for 5G wireless communication networks. *IEEE Communications Magazine*, 52(2), 122-130. DOI: 10.1109/mcom.2014.6736752
7. Popovsk, P., Brau, V., Mayer, H. P., Fertl, P., Ren, Z., Gonzales-Serrano, D., ... & Chatzikokolakis, K. (2013). EU FP7 INFSO-ICT-317669 METIS, D1. 1 *Scenarios, requirements and KPIs for 5G mobile and wireless system.*
8. Painuly, S., Kohli, P., Matta, P., & Sharma, S. (2020, December). Advance applications and future challenges of 5G IoT. In *2020 3rd International Conference on Intelligent Sustainable Systems (ICISS)* (pp. 1381-1384). IEEE. DOI: 10.1109/iciss49785.2020.9316004
9. Tudzarov, A., & Janevski, T. (2011). Design for 5G mobile network architecture. *International Journal of Communication Networks and Information Security*, 3(2), 112-123.
10. Son, H. J., & Yoo, C. (2015). E2E Network Slicing Key 5G technology: What is it? Why do we need it? How do we implement it?. Netmanias web page.
11. E2E Architecture Overview, available on: https://bit.ly/2SRd0u6
12. 3GPP SA2 architecture and functions for 5G mobile communication system. https://www.sciencedirect.com/science/article/pii/S240595951730 019X
13. Li, X., Samaka, M., Chan, H. A., Bhamare, D., Gupta, L., Guo, C., & Jain, R. (2017). Network slicing for 5G: Challenges and opportunities. *IEEE Internet Computing*, 21(5), 20-27. DOI: 10.1109/mic.2017.3481355
14. NFV, G. (2013). *Network Functions Virtualisation (NFV)*; Architectural Framework. NFV ISG.
15. Szabo, R., Kind, M., Westphal, F. J., Woesner, H., Jocha, D., & Csaszar, A. (2015). Elastic network functions: opportunities and challenges. *IEEE Network*, 29(3), 15-21. DOI: 10.1109/mnet.2015.7113220
16. Herrera, J. G., & Botero, J. F. (2016). Resource allocation in NFV: A comprehensive survey. *IEEE Transactions on Network and Service Management*, 13(3), 518-532. DOI: 10.1109/tnsm.2016.2598420

17. ETSI Industry Specification Group (ISG), (2013). Network functions virtualisation (nfv): Architectural framework. *ETsI Gs NFV*, 2(2), V1.

18. Szabo, R., Kind, M., Westphal, F. J., Woesner, H., Jocha, D., & Csaszar, A. (2015). Elastic network functions: opportunities and challenges. *IEEE Network*, 29(3), 15-21. DOI: 10.1109/mnet.2015.7113220

19. Mijumbi, R., Serrat, J., Gorricho, J. L., Bouten, N., De Turck, F., & Boutaba, R. (2015). Network function virtualization: State-of-the-art and research challenges. *IEEE Communications Surveys & Tutorials*, 18(1), 236-262. DOI: 10.1109/mcom.2009.5183468

20. Samdanis, K., & Taleb, T. (2020). The road beyond 5G: A vision and insight of the key technologies. *IEEE Network*, 34(2), 135-141. DOI: 10.1109/mnet.001.1900228

21. Alliance, N. G. M. N. (2015). 5G white paper. *Next generation mobile networks*, white paper, 1.

22. Wu, Q., Li, G. Y., Chen, W., Ng, D. W. K., & Schober, R. (2017). An overview of sustainable green 5G networks. *IEEE Wireless Communications*, 24(4), 72-80. DOI: 10.1109/mwc.2017.1600343

23. Becvar, Z., Rohlik, M., Mach, P., Vondra, M., Vanek, T., Puente, M. A., & Lobillo, F. (2017). Distributed architecture of 5G mobile networks for efficient computation management in mobile edge computing. *5G Radio Access Networks: Centralized RAN, Cloud-RAN and Virtualization of Small Cells*, 29.

24. Rimal, B. P., Van, D. P., & Maier, M. (2017). Mobile edge computing empowered fiber-wireless access networks in the 5G era. *IEEE Communications Magazine*, 55(2), 192-200. DOI: 10.1109/mcom.2017.1600156cm

25. Saini, H. K., Grover, J., & Khajanchi, M. (2019). Future Visions of Eco-Friendly 5G Communication: Exposing the Drivers. *Advances in Power Generation from Renewable Energy Sources* (APGRES).

26. Chakareski, J., Naqvi, S., Mastronarde, N., Xu, J., Afghah, F., & Razi, A. (2019). An energy efficient framework for UAV-assisted millimeter wave 5G heterogeneous cellular networks. *IEEE Transactions on Green Communications and Networking*, 3(1), 37-44. DOI: 10.1109/tgcn.2019.2892141

27. Agiwal, M., Saxena, N., & Roy, A. (2018). Ten commandments of emerging 5G networks. *Wireless Personal Communications*, 98(3), 2591-2621. DOI: 10.1007/s11277-017-4991-8

28. Yarrabothu, R. S. (2018). 5G: The Platform. In *Powering the Internet of Things With 5G Networks* (pp. 1-39). IGI Global. DOI: 10.4018/978-1-5225-2799-2

29. Singh, A. K., Srivastava, N., & Dixit, S. (2020). Optimizing Resource Allocation of MIMO-OFDM in 4G and Beyond Systems. In *Advances in VLSI, Communication, and Signal Processing* (pp. 241-249). Springer, Singapore. DOI: 10.1007/978-981-32-9775-3_24

30. Wei, Z., Yuan, J., Ng, D. W. K., Elkashlan, M., & Ding, Z. (2016). A Survey of Downlink Non-orthogonal Multiple Access for 5G Wireless Communication Networks, *arXiv: 1609.01856v1* [cs.IT] 7 Sep 2016.

31. Lim, Y.-G., Taehun Jung, Kim, K. S., & Chae, C.-B. (2017). Waveform multiplexing for 5G: A concept and 3D evaluation. 2017 European Conference on Networks and Communications (EuCNC). DOI: 10.1109/eucnc.2017.7980694

32. Xia, X., Xu, K., Wang, Y., & Xu, Y. (2018). A 5G-Enabling Technology: Benefits, Feasibility, and Limitations of In-Band Full-Duplex mMIMO. IEEE Vehicular Technology Magazine, 1-1. DOI: 10.1109/mvt.2018.2792198

33. Shehata, M., Elbanna, A., Musumeci, F., & Tornatore, M. (2018). Multiplexing gain and processing savings of 5G radio-access-network functional splits. *IEEE Transactions on Green Communications and Networking*, 2(4), 982-991. DOI: 10.1109/tgcn.2018.2869294

34. Zhang, L., Wu, Y., Li, W., Salehian, K., Lafleche, S., Wang, X., & Montalban, J. (2018). Layered-Division Multiplexing: An Enabling Technology for Multicast/Broadcast Service Delivery in 5G. *IEEE Communications Magazine*, 56(3), 82-90. DOI: 10.1109/mcom.2018.1700657

35. Sharma, S. K., Bogale, T. E., Le, L. B., Chatzinotas, S., Wang, X., & Ottersten, B. (2018). Dynamic Spectrum Sharing in 5G Wireless Networks with Full-Duplex Technology: Recent Advances and Research Challenges. *IEEE Communications Surveys & Tutorials*, 20(1), 674-707. DOI: 10.1109/comst.2017.2773628

36. Sarret, M. G., Berardinelli, G., Mahmood, N. H., & Mogensen, P. (2016). Can Full Duplex Boost Throughput and Delay of 5G Ultra-Dense Small Cell Networks? *2016 IEEE 83rd Vehicular Technology Conference (VTC Spring)*. DOI: 10.1109/vtcspring.2016.7504150

37. Lacy, A. M., Bravo, R., Otero-Piñeiro, A. M., Pena, R., De Lacy, F. B., Menchaca, R., & Balibrea, J.7 M. (2019). 5G-assisted telementored surgery. *British Journal of Surgery*, 106(12), 1576-1579. DOI: 10.1002/bjs.11364

38. Do, D. T., Nguyen, T. T. T., Le, C. B., & Lee, J. W. (2020). Two-way transmission for low-latency and high-reliability 5G cellular V2X communications. *Sensors*, 20(2), 386. DOI: 10.3390/s20020386

39. Qiao, J., Shen, X. S., Mark, J. W., Shen, Q., He, Y., & Lei, L. (2015). Enabling device-to-device communications in millimeter-wave 5G cellular networks. *IEEE Communications Magazine*, 53(1), 209-215. DOI: 10.1109/mcom.2015.7010536.

40. Udayakumar, E., & Krishnaveni, V. (2020). A Review on Interference Management in Millimeter-Wave MIMO Systems for Future 5G Networks. In *Innovations in Electrical and Electronics Engineering* (pp. 715-721). Springer, Singapore. DOI: 10.1007/978-981-15-2256-7_65

41. Abdelwahab, S., Hamdaoui, B., Guizani, M., & Znati, T. (2016). Network function virtualization in 5G. *IEEE Communications Magazine*, 54(4), 84-91. DOI: 10.1109/MCOM.2016.7452271

42. Brenes, J., Lagkas, T. D., Klonidis, D., Muñoz, R., Rommel, S., Landi, G., ... & Vilalta, R. (2020). Network slicing architecture for SDM and analog-radio-over-fiber-based 5G fronthaul networks. *Journal of Optical Communications and Networking*, 12(4), B33-B43. DOI: 10.1364/jocn.381912

43. Saha, R. K. (2019). Realization of licensed/unlicensed spectrum sharing using eICIC in indoor small cells for high spectral and energy efficiencies of 5G networks. *Energies*, 12(14), 2828. DOI: 10.3390/en12142828

44. Valenzuela-Valdés, J. F., Palomares, A., González-Macías, J. C., Valenzuela-Valdés, A., Padilla, P., & Luna-Valero, F. (2018, July). On the ultra-dense small cell deployment for 5G networks. In *2018 IEEE 5G World Forum (5GWF)* (pp. 369-372). IEEE. DOI: 10.1109/5gwf.2018.8516948

45. Naser Al-Falahy and Omar Y. Alani (2017). Technologies for 5G Networks: Challenges and Opportunities, *IEEE Computer Society*, 19(1), 12-20. DOI: 10.1109/mitp.2017.9

46. Hu, Y. C., Patel, M., Sabella, D., Sprecher, N., & Young, V. (2015). Mobile edge computing—A key technology towards 5G. ETSI white paper, 11(11), 1-16.

47. Aris, A. M., & Shabani, B. (2015). Sustainable power supply solutions for off-grid base stations. *Energies*, 8(10), 10904-10941. DOI: 10.3390/en81010904

48. Buzzi, S., Chih-Lin, I., Klein, T. E., Poor, H. V., Yang, C., & Zappone, A. (2016). A survey of energy-efficient techniques for 5G networks and challenges ahead. *IEEE Journal on Selected Areas in Communications*, 34(4), 697-709. DOI: 10.1109/JSAC.2016.2550338.

49. Rahman, J., Tahsin, A., & Ullah, S. E. (2018). Performance Analysis of a 5G Compatible Windowing and Overlapping Scheme Implemented Hybrid Precoded mmWave Massive MIMO NC-OFDM System. *American Journal of Electrical and Computer Engineering*, 2(2), 5-15.

50. Tarneberg, W., Karaca, M., Robertsson, A., Tufvesson, F., & Kihl, M. (2017, June). Utilizing massive MIMO for the tactile Internet: Advantages and trade-offs. In *2017 IEEE International Conference on Sensing, Communication and Networking (SECON Workshops)* (pp. 1-6). IEEE. DOI: 10.1109/SECONW.2017.8011041

51. Ge, X., Tu, S., Mao, G., Wang, C. X., & Han, T. (2016). 5G ultra-dense cellular networks. *IEEE Wireless Communications*, 23(1), 72-79.

52. Chen, M., Zhang, Y., Li, Y., Mao, S., & Leung, V. C. (2015). EMC: Emotion-aware mobile cloud computing in 5G. *IEEE Network*, 29(2), 32-38.

53. Li, C., Li, Y., Song, K., & Yang, L. (2016). Energy efficient design for multiuser downlink energy and uplink information transfer in 5G. *Science China Information Sciences*, 59(2), 1-8. DOI: 10.1007/s11432-015-5510-8

54. Kutscher, D. (2016, April). It's the network: Towards better security and transport performance in 5G. In *2016 IEEE Conference on Computer Communications Workshops (INFOCOM WKSHPS)* (pp. 656-661). IEEE.

55. Panwar, N., Sharma, S., & Singh, A. K. (2016). A survey on 5G: The next generation of mobile communication. *Physical Communication*, 18, 64-84. DOI: 10.1016/j.phycom.2015.10.006

56. Rost, P., Banchs, A., Berberana, I., Breitbach, M., Doll, M., Droste, H., & Sayadi, B. (2016). Mobile network architecture evolution toward 5G. *IEEE Communications Magazine*, 54(5), 84-91. DOI: 10.1109/mcom.2016.7470940

57. Sridhar, S., & Smys, S. (2017, January). Intelligent security framework for iot devices cryptography based end-to-end security architecture. In *2017 International Conference on Inventive Systems and Control (ICISC)* (pp. 1-5). IEEE. DOI: 10.1109/icisc.2017.8068718

58. Maroc, S., & Zhang, J. (2019, July). Comparative analysis of cloud security classifications, taxonomies, and ontologies. In *Proceedings of the 2019 International Conference on Artificial Intelligence and Computer Science* (pp. 666-672). DOI: 10.1145/3349341.3349487

59. Patzold, M. (2018). 5G readiness on the horizon [mobile radio]. *IEEE Vehicular Technology Magazine*, 13(1), 6-13.

60. Weber, M., & Boban, M. (2016, May). Security challenges of the internet of things. In *2016 39th International Convention on Information and Communication Technology, Electronics and Microelectronics (MIPRO)* (pp. 638-643). IEEE.

61. Al-Falahy, N., & Alani, O. Y. (2017). Technologies for 5G Networks: Challenges and Opportunities. *IT Professional*, 19(1), 12-20. DOI: 10.1109/mitp.2017.9

62. 5G Reuse cases and requirements, Nokia Networks, white paper. Available at: http://networks.nokia.com/sites/1042/default/files/document/5g_requirements_white_paper.pdf.

63. Lee, C. N., Lee, M. F., Wu, J. M., & Chang, W. C. (2018, November). A Feasible 5G Cloud-RAN Architecture with Network Slicing Functionality. In *2018 Asia-Pacific Signal and Information Processing Association Annual Summit and Conference (APSIPA ASC)* (pp. 442-449). IEEE.

64. Omheni, N., Bouabidi, I., Gharsallah, A., Zarai, F., & Obaidat, M. S. (2018). Smart mobility management in 5G heterogeneous networks. *IET Networks*, 7(3), 119-128. DOI: 10.1049/iet-net.2017.0208

65. Ding, A. Y., & Janssen, M. (2018). Opportunities for applications using 5G networks. *Proceedings of the Seventh International Conference on Telecommunications and Remote Sensing - ICTRS'18*. DOI: 10.1145/3278161.3278166

66. Benevoloetal Smart Mobility in Smart City in Empowering Organizations 2016. 13-28.

67. Jordaan, C. G., Malekian, N., & Malekian, R. (2019). Internet of things and 5G solutions for development of smart cities and connected systems. *Communications of the CCISA*, 25(2), 1-16.

68. Chiariotti, F., Condoluci, M., Mahmoodi, T., & Zanella, A. (2017). SymbioCity: Smart cities for smarter networks. *Transactions on Emerging Telecommunications Technologies*, 29(1), e3206. DOI: 10.1002/ett.3206

69. Aljohani M., & Alam T. (2017). Real Time Face Detection in Ad Hoc Network of Android Smart Devices. In: Sahana S., Saha S. (eds) *Advances in Computational Intelligence*. Advances in Intelligent Systems and Computing, vol 509, 245-255. Springer, Singapore. DOI: 10.1007/978-981-10-2525-9_24

70. Alam, T., & Aljohani, M. (2015, November). An approach to secure communication in mobile ad-hoc networks of Android devices. In *2015 International Conference on Intelligent Informatics and Biomedical Sciences (ICIIBMS)* (pp. 371-375). IEEE. DOI: 10.1109/iciibms.2015.7439466

71. Orlosky, J., Kiyokawa, K., & Takemura, H. (2017). Virtual and Augmented Reality on the 5G Highway. *Journal of Information Processing*, 25(0), 133-141. DOI: 10.2197/ipsjjip.25.133

72. ABI Research and Qualcomm: *Augmented and Virtual Reality: The First Wave of 5G Killer Apps*. White paper (2017). Online available: https://bit.ly/2LhoAe0

73. Yu, H., Lee, H., & Jeon, H. (2017). What is 5G? Emerging 5G mobile services and network requirements. *Sustainability*, 9(10), 1848. https://bit.ly/3fERUJo

74. Ullah, H., Nair, N. G., Moore, A., Nugent, C., Muschamp, P., & Cuevas, M. (2019). 5G Communication: An Overview of Vehicle-to-Everything, Drones, and Healthcare Use-cases. *IEEE Access*, 7, 37251-37268. DOI: 10.1109/access.2019.2905347

75. French, A. M., & Shim, J. P. (2016). The digital revolution: Internet of Things, 5G, and Beyond. *Communications of the Association for Information Systems*, 38(1), 840-850. DOI: 10.17705/1cais.03840

76. Latif, S., Qadir, J., Farooq, S., & Imran, M. (2017). How 5G Wireless (and Concomitant Technologies) Will Revolutionize Healthcare? *Future Internet*, 9(4), 93. DOI: 10.3390/fi9040093

77. Framework and Overall Objectives of the Future Development of IMT for 2020 and Beyond, ITU-R M.2083, 2015.

78. Sung, M., Cho, S.-H., Kim, J., Lee, J. K., Lee, J. H., & Chung, H. S. (2018). Demonstration of IFoF-Based Mobile Fronthaul in 5G Prototype With 28-GHz Millimeter wave. *Journal of Lightwave Technology*, 36(2), 601-609. DOI: 10.1109/jlt.2017.2763156

79. Gaur, L., Afaq, A., Solanki, A., Singh, G., Sharma, S., Jhanjhi, N. Z., ... & Le, D. N. (2021). Capitalizing on big data and revolutionary 5G technology: Extracting and visualizing ratings and reviews of global chain hotels. *Computers & Electrical Engineering*, 95, 107374.

80. Ge, C., Wang, N., Selinis, I., Cahill, J., Kavanagh, M., Liolis, K., & Poziopoulou, G. (2019). QoE-Assured Live Streaming via Satellite Backhaul in 5G Networks. *IEEE Transactions on Broadcasting*, 1-11. DOI: 10.1109/tbc.2019.2901397

2

Scope and Challenges of IoT and Blockchain Integration

Sara Shree[1], Monika Sharma[1]

[1] AIIT Amity University, U.P., India
Email: sarashree379@gmail.com, msharma5@amity.edu

Abstract
The internet of things (IoT) is a somewhat new and booming technology that is somewhat automated in nature, which is extending itself day by day by various means and sources. It has a colossal ability to reach an intelligent level of connectivity by numerous means. The IoT has the ability to connect billions of devices and physical objects at once, which enhances the sharing of information on a digital basis. Although the IoT has unlimited advantages, it also comes with several issues due to its centralized system known as the server-client model. This chapter presents a significant study on integrating two major technologies: blockchain and the IoT. It discusses how blockchain can play a key role in solving several issues and how several domains and platforms of blockchain can act as an update provider to IoT services and its assets. It also focuses on the pros and cons of the integration of both the technology and the existing platforms based on the alliance of IoT and blockchain platforms like Ethereum, Hyperledger, Lisk, and Slock.it, which are explained along with their full funtionality.

Keywords: Internet of things, blockchain, automated technology, centralized systems, Ethereum, Hyperledger, Slock.it, integration of blockchain and iot

2.1 Introduction

As human society continues to advance in the modern world, there continues to be exponential development in the modern electronic system. Wireless communication has played an important role in developing a new generation of technology. It has produced positive results that have increased suitable electronic devices for so many required areas, and has also boosted the production cost and the quality of the products, which are producing positive results in correspondence to the new technologies. The emerging IoT technology has unfolded in a set of technologies, such as wireless sensor network (WSN) and radio frequency identification (RFID), whose accelerated mechanism are sensor based, which allows them to communicate across the internet.

Currently, the IoT is used as an electronic device for smartwatches to smart shoes to calculate the number of steps walked. The IoT expresses itself in various parts of society's development format and has a range of applications which complete several general and personal needs of society. It has played a mid-way role in the development of smart cities, turning normal houses into smart houses by providing automation to the electric grid system. Various research has shown that the number of connected devices will be 22 to 50 million by 2021. The IoT is a live visualization of a connected world where data plays a very important role in the communication between the devices. With the huge expansion of the new IoT technology, IoT comes with several issues. The IoT runs on a sequential mechanism and multiple protocols that helps to reduce the multifariousness in its regular functioning but the complex multifariousness in new IoT technology is reducing its adaptability in some areas of implementation.

Apart from this problem, the biggest issue occurring day by day is IoT data being compromised each day due to its centralized system. The IoT is being used by both government and private sectors, which store personal information by various means and for various purposes. The centralized IoT technology has recurring flaws and the personal information included in various functions of the IoT can be compromised easily. There are several solutions that can provide security to IoT services. Blockchain is one of these solutions, which not only provides security but also enhances the current IoT systems.

Infrastructure protocols of blockchain technology are a solution for issues in the IoT. Bitcoin is supported by a protocol infrastructure that ensures that the required information is stored and can be used when needed, like a ledger which stores information that can be obtained only when required. Blockchain technology protocol infrastructure allows a transaction to be verified by several blocks present in the blockchain.

The revolutionary assurance factor that blockchain has developed has made it useful in the critical private and government sectors, which are adopting IoT technologies.

In this chapter, we will cover the alliance of IoT and blockchain, including the advantages and disadvantages of this alliance between private and government sectors. The centralized system of IoT and Blockchain previously described in this chapter are very important in terms of understanding the role of blockchain in IoT. There are various blockchain platforms for IoT and various applications that are useful from the integration of this technology.

2.2 Literature Review

Dorri *et al.* [1] have proposed an IoT architecture which is lightweight and totally new and mainly focuses on security and privacy. This is totally based on blockchain technology.

The example of smart homes was shown with the help of this architecture, which neglects the weight of the IoT with the help of blockchain and maintains critical functions of smart homes, including security and privacy features in the background. The suggested planning is hierarchical, with an overlay network that consists of the juncture of a smart home; cloud storage is used for data transaction.

Gupta *et al.* [2] have proposed a positive approach to a secure healthcare system with the help of blockchain, which will provide a secure and safe health record system in which consumers are the owners of their own records. Their system proposes a mechanism for storing just metadata from the records (includes personal data) in the blockchain. Furthermore, they explain the main advantage of this system in keeping the digital records in the healthcare industry at a very large scale with the help of blockchain. The real data and personal information are stored in a universal cloud and consumers have access to the data they are the owner of. And information that plays a key role in their treatment is stored in blockchain information like patient identity, visit id, and provider id.

Reyna *et al.* [3] did a complete analysis of the issues which will arise on the integration of IoT and blockchain. They also present several methods to integrate both the platforms and how they can be applied at an industrial level. They categorize their work by exploring IoT systems like smart home and smart car, and how blockchain can be applied to them to enhance their security. They also explain several marketplaces for the integration of IoT and blockchain and which types of opportunities they can generate at an industrial level.

Conoscenti *et al.* [4] explain several IoT security concerns in detail and how blockchain can solve the issue of its flaws being exploited by using blockchain as countermeasures. Furthermore, the integration of both technologies is presented, and how it can not only provide various security options but can also help several IoT services to run flawlessly.

Christidis and Devetsikiotis [5] classify several blockchain technologies and their domains and analyze the advantages and disadvantages of blockchain being introduced to IoT. They propose different sets of ideas like the use of a blockchain and interplanetary file system for regularly updating IoT services. This paper also describes the use of smart contracts and how they are set up in the marketplace by using various IoT platforms where IoT services can earn money by providing their resources.

Jentzsch [6] describe the first-ever IoT blockchain integrated platform known as Slock.it, which was developed by a blockchain platform known as Ethereum. The functionality of this system represents how real-world physical objects can be controlled by blockchain and IoT integrated services. This paper also explained how smart services can be rented for usage and payments can be done directly without the use of a third party.

2.3 Internet of Things and Its Centralized System

Kevin Ashton, MIT's Executive Director of Auto-ID Labs, coined the phrase "Internet of Things" in 1999, although it took at least another decade for the technology to catch up with the vision. In simple terms, it can be defined as objects (physical) or the set of devices, and the set can be defined as any required smart network that can be of the building (smart home), the vehicle (smart vehicle) and various other sets; and in the IoT, these devices interchange data with the help smart circuits, wires, software and connectivity tools, and these systems are fully automated so that the IoT is totally independent of humans for their functions.

Moreover, IoT technology extends to perform and add functions of data exchange in automatic technology like automation data transfers, home automation, robotics, and many more. The most recent IoT technology to have aggressively evolved are wireless sensor network (WSN), smart barcodes, intelligent sensing, RFID, NFCS, and cloud data storage.

The current IoT system involves a central client/server system which is used for identification and keeps all the devices connected; however, this centralized system of IoT restricts its widespread use. Decentralizing the existing IoT system will be the precise decision that will help to increase the reach of IoT in many ways. That's where blockchain plays an important role.

In IoT, different devices are connected to the internet and they communicate by various automated sensors and data analyzed from the IoT ecosystem. There are various methods through which these devices maintain the connection or get connected to the internet. The methods can be networking nodes, servers, or smart and normal computers. Several researchers have given various architecture models for IoT, but there is a common IoT architecture which is approved and proposed by ITU.

The centralized IoT architecture is composed of three layers, as shown in Figure 2.1.

- Application layer and service support
- Network layer
- Device layer

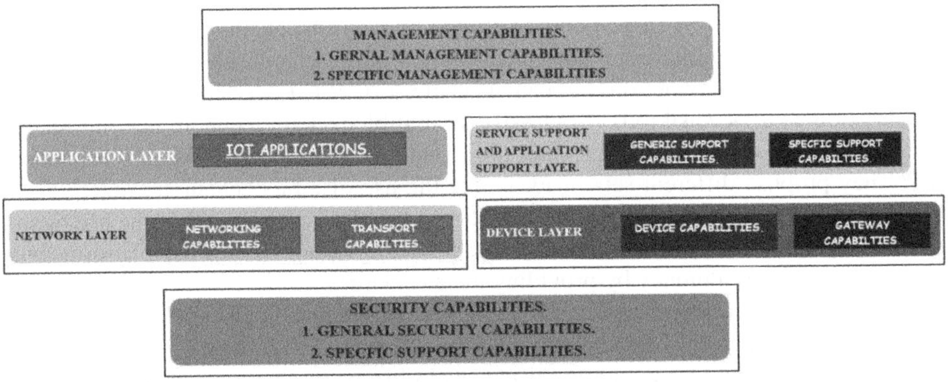

Figure 2.1: IoT refrence model and architecture.

2.3.1 Application Layer

The application layer contains relevant IoT applications commonly used in the functioning of applications. This layer is at the forefront of various automated programs like healthcare, smart cities, connected cars, and smart electrical grid system. All the bigger infrastructure of IoT goes through the application layer.

2.3.2 Network Layer

As the name suggests, the network layer is used to provide internet connectivity services to the local area and wide area networks such as routers, firewalls, and gateways. It also provides a platform for interconnection communication between the smart devices which can be further used as a remote control for smart devices such as smartphones. Every device which is connected through IoT requires internet connectivity and the work of the network layer is to provide that connectivity.

2.3.3 Device Layer

In an open system interconnection (OSI) model, the device is similar to the OSI model architecture of the network. Control of the smart object is handled in the device layer by the controllers. All the smart objects in the device justify their connection through several endpoints. Devices receive and send a variety of information. The current IoT framework is a centralized server/client model in which all devices communicate through the centralized system; however, this restricts the widespread large-scale use of the IoT.

A decentralized system will play an important role in the expansion of the IoT and will secure several of the IoT services which are vulnerable in a centralized system. The decentralized system will not only ensure the IoT's security, it will bring various useful changes to the IoT architecture.

2.4 Blockchain Technology

Blockchain technology is a recent technology which provides efficiency in securing online transactions (it can be of any type); in technical terms we can say it is a secure framework for digital interconnections. However, since this new technology has only recently been introduced on the market, it is being adapted day by day for use as stand-alone technology or integrated with other technologies like IoT and artificial intelligence.

An exact definition of blockchain does not exist. But in simple and precise terms we can say it is a technology that keeps records in the form of a ledger with the help of databases of any digital interactions between two or more parties, devices, servers, etc. Each transaction is verified by the people involved in the interaction and by other frameworks of blockchain.

There are two important elements in blockchain:

1. Transactions: Important actions that are generated by the participants in the interaction of the blockchain environment.

2. Blocks: Blocks ensure that records of the transactions are kept in a correct and sequential manner in the form of a ledger.

2.4.1 Characteristics of Blockchain

Blockchain has may attractive features which will help to resolve several problems of the IoT services. These features are listed below and shown in Figure 2.2.

- Immutability: The key feature of blockchain is an immutable ledger. Databases and services which are centralized are prone to various data theft and attacks if one node of the system is down, which can bring down the whole centralized system that is totally dependent on a third for the security of the information. But the decentralized system of the blockchain includes a special immutability feature which ensures that once the transaction is recorded and approved it can't be tampered with or revert back.

- Decentralization: Blockchain decentralizes everything. There is no single center support in the blockchain and every node in the network is totally independent, so a failure in one node doesn't compromise the whole system. Even though a centralized system assures scalability and strength by utilizing the resources from every single participating node, the chances of failure are greater in a centralized system.

- Anonymity: Blockchain system provides anonymity to every single user pre- and post-transaction. It has a unique key for every single participant in the interaction which is not easily accessible by any means, so it provides complete security to data and the information on the transactions that have been done in the digital interface through blockchain.
- Increased Capacity: The most attractive feature of blockchain is to provide increased capacity from normal storage in the centralized systems. Thousands of computers and servers working together are more powerful than a centralized system.

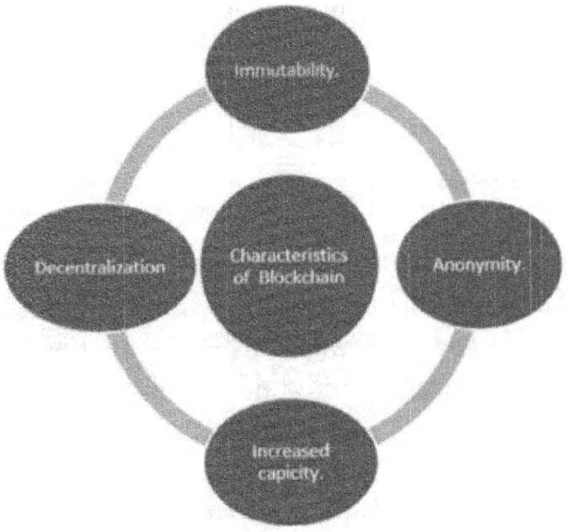

Figure 2.2: Characteristics of blockchain.

2.4.2 Working Mechanisms of Blockchain

Being a new technology, blockchain continues to evolve day by day and as it is evolving it is solving all the problems of modern technologies such as important data-driven decisions, data theft issues, and anonymity issues.

For storing all the information, blockchain uses several blocks which are recorded in databases with various assurances like immutability, anonymity, data protection, and many more. When the new transaction is performed in the blockchain the sender notifies every single node in the blockchain through disposals of various communication channels. For the validation of the transaction, nodes keep it in the ledger as a record.

The validation of the transaction is run by performing pre-planned checks on the actions of the transactions. Miners are special nodes in the network which collect some or all the available data from the transaction pools. Block header plays an important role in finding the proper proof of the transaction, and for this the whole block is mined and the variable data is extracted, which will be used for determining the authenticity of the transactions. In other words, this whole process is cryptographic in nature and runs on continuous calculations and data transfers. Since mining requires a lot of processing, the individuals who perform the mining process use dedicated software.

Every new block created has a timestamp that is shared between all the nodes in the network. After every node receives the block and its transaction is validated, that block is added to the ledger. When a high number of nodes receive the blocks in the network it becomes an unchangeable part of the blockchain. Every block stores a metadata value of the previous block so every block has a signature of the previous block. That's how all the blocks are linked together, creating a chain called a blockchain.

2.4.3 Example of Blockchain Transactions

We have four nodes (*W*, *X*, *Y*, *Z*) that want to transfer the cryptocurrency bitcoin using the blockchain technology. In this process, there will be no intermediate third party to help in the transaction process, which is the idea of decentralization. Therefore, there will be a direct transfer of bitcoin between *W* and *X*.

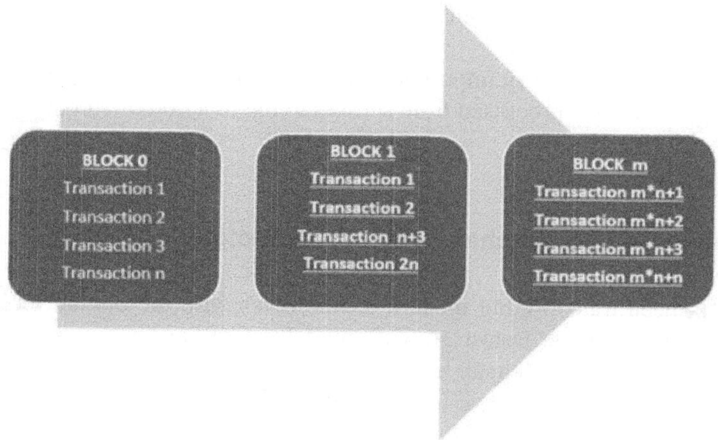

Figure 2.3: Example of blockchain technology.

In Figure 2.3, we can see that if *W* wants to send 10 bitcoins to *X* then an individual transaction is created which is verified by each node; in the same manner, if *X* wants to send 20 bitcoins to *Y*, again each transaction is verified by each node in the ledger. All the transactions are chained together in what is called a distributed ledger because the recorded ledger is distributed across all the nodes in the network, which is taken as a validation for the transactions.

2.4.4 Need for Blockchain Technology

Various sources have accepted the fact that the IoT has many vulnerabilities. Today, there are 5 billion devices which are connected to the IoT and the numbers are predicted to grow by 29 million to 75 million by 2023. From these numbers we can deduce that there will be a continuous production and flow of data for IoT functions. Therefore, some fundamental issues need to be addressed before moving on to the bigger issues of theft and security because every big issue is related to small vulnerabilities. In this section, we will discuss several IoT issues and blockchain as a solution to those problems.

One issue that the IoT faces with a distributed architecture with a centralized system is that every node in the process can be easily compromised. A common attack on these systems is distributed denial of service (DDoS); and in a centralized distributive network if multiple devices are compromised then it can easily shut down all the systems.

Here, blockchain plays an important role in decentralizing this centralized IoT technology, making every node functionality independent. There is a very smooth mechanism of verification of every digital interaction which happens in the network inside the node.

The IoT technology is widely used in various sectors of private and government services. Therefore, because this technology is used so much, the issue of its vulnerability to attacks can cost time, money, and everything else connected to IoT services.

Another issue is the integrity of data in IoT technology [7]. Since the IoT totally runs on decision support systems, a timely decision is made with the help of the analyzed and processed data from the sensors. Thus, it is very important to secure the system from the insertion of the injection attacks which will falsify the decision-making framework of IoT. For proper functioning of automated IoT systems, such as vehicular networks, automated manufacturing industries, and smart grids that process data in real time, which make decisions based on data from sensors, the downtime of sensors can result in critical situations.

Blockchain can play a very important role in providing security in terms of end-to-end encryption while data is processed from sensors to the function of automation objects.

2.5 Integration of Blockchain and IoT Technology

The IoT has been transforming modern technologies with the optimal manual capacity to make them part of the fastest growing models of the technological era. Processing and obtaining data at meta-human levels, this technology changes the picture by inventing various fast pathways which have improved the management and quality of social life in terms of digitalization.

Currently, the functionality of IoT technology has been supported by cloud computing for analyzing and processing data which is used for real-time processing.

Cloud computing is a centralized way of providing services with a single framework acting as administrative support. Centralized architecture has so many vulnerabilities and issues which compromise data in terms of transparency. Participants in this process of using IoT services have no information on how securely their submitted data is being processed for digital interaction.

As previously discussed in prior sections of this chapter, we can see that two promising integrated technologies, cloud computing and the IoT, are vulnerable to so many threats.

Blockchain can help IoT in providing sharing services that can be trusted. Information can be provided in a reliable way and can be traced for its secure packet to travel through the network. The original source of data can be identified and cannot be changed or tampered with, which enhances its security.

For instance, in a food distribution system where the IoT is used to ensure security by monitoring and tracing several food packets, the proper function of traceability of the data must be shared between trusted persons who are sponsoring and managing the food safety events. Distribution of food and securing it involves many steps like manufacturing, treatment, distribution, and feeding; and remembering all these events are monitored in a digital manner and if there is a vulnerability at any point then it will compromise the whole automated chain of the food distribution system which is powered by the IoT. Here, if data is manipulated, the manipulated wrong information can easily be provided in place of the correct information, which will disrupt the whole automated food distribution system, which will affect so many things and will compromise so many lives directly and indirectly.

The use of blockchain technology in modern IoT technologies is not just reliable, it also provides various security features, which in scenarios like food distribution will come in handy for providing high-end security features like personal key verification for every person who is accessing the databases and end-to-end insertable encryptions. Blockchain will not only help solve several issues concerning the IoT, it will also play an important role in improving the IoT technology while they are being integrated.

But this new technology also needs improvements. It doesn't matter whether the new technology on the market is solo or integrated, it will always come with several advantages and disadvantages. In the coming sections, we will look at how these two technologies integrate themselves.

2.5.1 Interactions of IoT and Blockchain Integrations

The two major technologies of the IoT and blockchain are integrated for solving various issues of the IoT. For proper interactions, the blockchain should have access to the underlying layers of IoT which are directly rooted in the centralized system of the IoT. In addition to this, there are vast possibilities of how IoT can be integrated with blockchain.

There are mainly three approaches to how IoT can be integrated with the blockchain technology (see Figure 2.4):

1. IoT-IoT
2. IoT-Blockchain
3. Hybrid approach

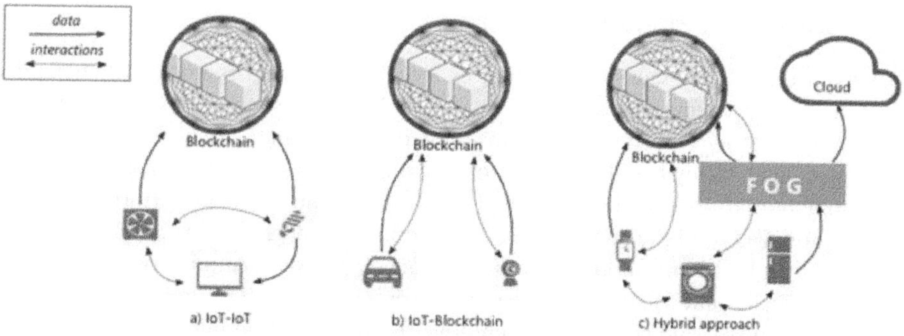

Figure 2.4: Interactions between blockchain and IoT.

2.5.1.1 IoT-IoT

This interaction of IoT-IoT could be the fastest for security and fast interaction between the IoT and several IoT devices. IoT devices are required to discover important nodes for communications, which is known as a routing mechanism. Here the blockchain will be used for data storage and part of the data will be stored in the blockchain. This approach is reliable and secure for local digital transactions of the data and can be used in the transfer of big bunches of the data blockchain in the background, which will keep the record in the form of a ledger.

2.5.1.2 IoT-Blockchain

This interaction involving blockchain will act as a gateway for every interaction, which will enable the permanent record of the interaction. Every interaction in this type will ensure the traceability of the data and all data is stored in the form of a ledger, which also ensures the anonymity of the data. Recording every single transaction will require a larger infrastructure, larger bandwidth, and more data interpretation tools, which will be a challenge to these two integrating technologies.

2.5.1.3 Hybrid Approach

In a hybrid approach, there would be a direct sharing of IoT devices, with only a part of the data and interaction taking place in IoT. The biggest challenge in this interaction is choosing importation interactions which will pass through blockchain and a run-time also should be decided for them. This technology is totally based on the real-time scenario so both technologies can be used in the best possible way; therefore, we can benefit from the integration of both the technologies. Moreover, the hybrid interaction will provide a lot of benefits and improvements to IoT technology.

Devices are used in a very limited manner in the IoT deployment. Resources are extracted from the end nodes and they have the responsibility of forwarding the data collected from the sensors to the upper layers. Blockchain integration with IoT comes with certain advantages like there being various keys generated for the purpose of anonymity, which are provided to the gateways of blockchain. These keys are plausible in terms of traditional deployment. But the deployment of gateways comes with few benefits because databases are used more frequently for executing the applications and tasks. However, an unsecured database comes with a lot of vulnerabilities and will cost the same amount of money and time that can be used to deploy hybrid and several other approaches to secure the framework of the IoT technology.

2.5.2 Blockchain Platforms for IoT

Blockchain technology can be identified as a technology that is not constant; and even though it keeps evolving and changing on a daily basis, it continues to have a great effect on modern industries. Since the platforms related to blockchain are high in number, all the platforms cannot be analyzed and monitored. In this section, we will discuss the popular and adaptable platforms of blockchain for IoT. Ethereum [8] is a platform where smart contracts are used with the support of the blockchain. Ethereum runs on the functionality of blockchain with a built-in programming language (solidity). Another built-in feature of a virtual machine is called an Ethereum virtual machine (EVM). This integration provides an opportunity to invest and adopt this technology as an integration into the other technology. Currently, Ethereum is used as the most popular format for developing applications for IoT. The definition of smart contracts can be told in a simple manner where companies and their products can publish their policies and measures taken to react to certain changes.

There is an open-source platform for blockchain-related development known as Hyperledger. In Hyperledger, there is a fabric known as a Hyperledger Fabric, a blockchain with sets of permission that don't support cryptocurrency but has platform support for commercial implementations like IBM's blockchain platforms [9]. It also provides a platform for online membership and consensus using different components from the platforms. By using general-purpose computer languages this integrated platform can be used to build distributive applications. IBM's Bluemix platform eases the integrations of IBM's blockchain with other technologies. Food traceability with IoT is a project run on this platform.

There is an underdeveloped platform named HDAC, which is an IoT contract with M2M interaction platform based on blockchain. The HDAC systems use a public and private blockchain network and tag every transaction with a quantum number for more secure interactions. This platform should be more widespread and open for the public and private sectors because of its advanced security features.

Lisk is another platform for IoT-integrated blockchain technology. It has a decentralized system with sidechains into the integrated platforms which give the choice of using multiple cryptocurrencies or multiple integrated cryptocurrency systems such as bitcoin and Ethereum. Lisk creates an adaptable environment for blockchain services, and end-users directly use this platform because of its ability to create and deploy decentralized distributive applications. Currently, Lisk technology is integrated with IoT's chain of things to look up the possibility of extended security which can be provided to the IoT.

Litecoin [10] platform is similar to bitcoin but is a lot faster; the transaction speed in bitcoin is 10 minutes and the transaction speed of Litecoin is 2.5 minutes. Proof of the transaction provided by Scrypt, a security feature based on an intensive password key. Litecoin also requires less computational power to operate and hence also results in fewer nodes being used, which makes this platform more efficient and suitable for integration with IoT.

The Quorum [11] platform was developed for privacy in financial industries. The implementation of this platform is done with the use of Ethereum, which provides secure transactions with the help of permissions when the transactions are in transit. Cryptographic methods are used for data privacy and contract privacy. Recently, this platform was integrated with Zerocash technology to observe and attain all the information on the transactions which are taking place in the live window.

2.5.3 Advantages of Integrating IoT with Blockchain

The advantages of integrating IoT with blockchain are described below and shown in Figure 2.5:

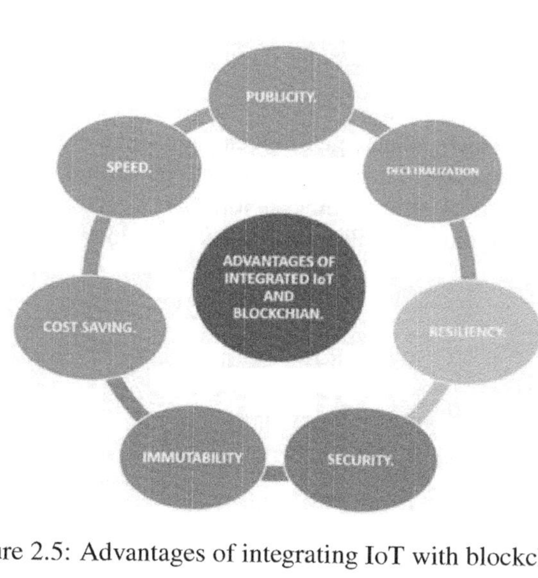

Figure 2.5: Advantages of integrating IoT with blockchain.

- Publicity: Blockchain has an individual block that stores every transaction in the form of a ledger so there is total transparency in the interactions and the participants who are in the interaction. Each participant has an individual private key which provides a secure and private way to prevent the data breaches and integrity of the privacy.

- Decentalization: To add the transaction into the interface it must be verified by the participants for their approval before being added to the ledger. No single authority approves the transactions. There is just one rule, which is that every participant in the interface has to approve the transactions. This generates a massive amount of trust in the participants and the system interface [12]. This system will come in handy for use in IoT because IoT functions on a centralized system and a single point of failure can bring down the entire system; however, the decentralized system will help to secure the IoT from failures and threats.

- Resiliency: Each node has its data copied in its own ledger which contains all the transactions which were made in the digital interface of the blockchain network. So, blockchain has the ability to stand against various cyber threats, and even if a single node is compromised the whole system will be up and running and the threat can be neutralized in the background. Information sharing needs will be improved in IoT systems, but it raises other issues like processing and storage.

- Security: IoT runs on numerous untrusted parties; in other words, it is a heterogeneous network. Blockchain has the ability to provide security from untrusted parties; simply put, all IoT nodes should be able to withhold any kind of malicious attack.

- Speed: The distribution of transactions in the blockchain is very fast across the network and can be processed anytime and anywhere. The integration of IoT and blockchain will bring this ability to the IoT services and will also speed up the IoT frameworks with better interaction. Moreover, due to the blockchain's decentralized system, the dependency on centralized systems will not be a big issue.

- Cost Saving: Due to the centralized structure, IoT requires high maintenance and infrastructure that includes large server farms, communication, and networking equipment which is expensive, and the cost will keep increasing more and more as IoT devices continue to be added day by day. Blockchain is a correct solution for the centralized structure of IoT.

- Immutability: The main advantage of blockchain technology is an immutable ledger. Once the ledger saves the data which is verified by several sources, altering and tampering with the data is almost impossible [13,14]. As a result of these integrated technologies, the immutable ledger will provide a framework for several IoT services.

2.5.4 Challenges of Blockchain and IoT Integration

The challenges of blockchain and IoT integration are given below and shown in Figure 2.6:

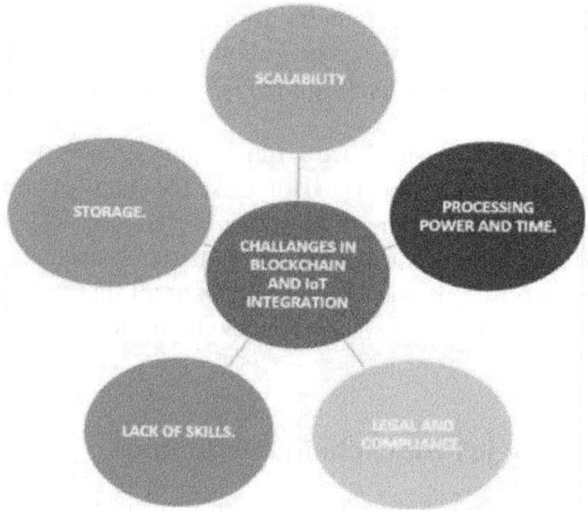

Figure 2.6: Challenges of blockchain and IoT integration.

- Scalability: The peculiar function of blockchain requires it to function on higher nodes, due to which a scalability dilemma arises; and if this issue remains in the picture then blockchain technology will also move towards the centralized composition. This is a very alarming and serious issue because IoT networks contain a substantial number of nodes and the integration will nearly come to a standstill.

- Processing Power and Time: Power and processing time is needed to encrypt all the devices with blockchain. IoT systems have multiple kinds of devices that run on the multiple different configurations with different computing capabilities, all of which can't run on the same algorithm at the required speed. This configuration of IoT calls for securing IoT with blockchain. Blockchain has to adapt according to every configuration that is configured for the devices in IoT. This will increase the applied processing power and storage, and ultimately, it will increase the cost of this integration. For the purpose of cost reduction, this integration is placed in the first place only.

- Storage: The main benefit of blockchain is that it is a decentralized system, which negates the need for a central system to record the transactions and device id's. The ledger of the blockchain is stored in the node itself [15]. With the increasing number of nodes for distributive ledger, the ledger sizes will also increase for IoT nodes because the blockchain will use the IoT nodes only in a more secure and hybrid integration. As we have explored earlier, IoT runs on less computational resources and storage capacity [16].

- Lack of Skills: The blockchain technology is new and it keeps changing and readjusting. So, there are very few people with complete knowledge and skills in blockchain and its domains, especially the growing domain of cryptocurrency. There are a large number of people who don't understand how blockchain works. On the other hand,

IoT technology exists everywhere, so it is very difficult to deploy the integration of both technologies without public awareness.

- Legal and Compliance: Being a new technology, blockchain has many abilities that help to the other new technologies in the form of integration around the globe. Moreover, since its usage will increase the echoing of its presence, its capabilities will increase, making it even more useful in the developments of mankind.

But since this technology doesn't have a functionality code to follow, there is no governing body for it, which raises the issue of trust for many manufacturers and service providers. Take the bitcoin blockchain-powered cryptocurrency, for example, which is banned in many countries due to various issues that sound more alarming when criminals use bitcoin for their purposes. If we take into account all that we know, this problem can act as a major barrier for the integration of blockchain, not just with IoT but with every other new technology.

2.5.5 Applications of IoT-Blockchain Integration

The applications of IoT-Blockchain integration are given below and summarized in Table 2.1.

Table 2.1: IoT-blockchain applications.

Application	Classification	Platform
LO3 Energy	Energy microgrid	Ethereum
Aigang	Insurance network for IoT assets	Ethereum
MyBit	Investment in IoT devices	Ethereum
Aerotoken	Sharing airspace market for drone navigation	Ethereum
Chain of Things	Identity, security	Ethereum
Chronicled	Identity, data provenace and automation	Multiplatform
The Modum	Data integrity for the supply chain	Multiplatform
Twin of Things	Sharing and machine economy	Multiplatform
Blockchain of Things	Secure connectivity between IoT devices	Multiplatrorm

- LO3 Energy [17] is an energy microgrid that is being used in Brooklyn (USA), which is assisted by blockchain system for its functioning. Other countries, like Germany and Australia, also use LO3 Energy. Storing electrical energy and generating it in a localized group is referred to as a microgrid. Its main function is to establish coordination with a broader power grid with the help of P2P decentralized systems. This system is developed for a community marketplace to support other projects with the same mechanism. It is the first blockchain platform for energy-based usage. It comes with enhanced security allowance for direct transit of energy sales among the participants, by allowing the device at the edge of the grid. The measurement of energy production is done by a hybrid device that is placed where LO3 provides services at a localized level and data is collected which is forwarded in the network for observations.

- Aigang: A special autonomous network that provides insurance for IoT assets, Aigang uses Ethereum to issue policies for the assigning of smart contracts. The use of Ethereum here provides security because it performs risk assessments and claims of the insurance automatically. Aigang has its own virtual currency known as AIX. Aigang offers several investment options at a certain risk level while providing potential gains on the other hand. The main aim of this integrated tech is to provide insurance policies with automation with the help of smart contracts.

- MyBit [18]: The aim of MyBit is to build an ecosystem of data-sharing services. Several people can share their owned IoT assets (e.g., drones, smart cars). A new investment option with an cutting-edge financial model is open to the public. Ethereum is used here for several automation processes. When these IoT assets do well in the marketplace, the owners receive their gains according to the ownership stake they have on several IoT assets. For monitoring all the ecosystems in a proper sequential manner, a centralized system is used. The platform justifies different IoT asset types and several IoT devices are linked to the assets; after the installation they send and receive information through various APIs. Oracle is used to connect devices to the network.

- Aerotoken [19]: This system is created for low altitude commercial drones. The system is used in real-time automated navigation and property-access authorization. It is a solution for accessing properties via drone in a shared airspace environment. It creates a lot of marketplace opportunities in shared environments and also solves the major problem of getting permission for drone operations. The owners of the property provide their respective airspace through smart contracts through blockchain; owners are paid on a temporary basis only for providing the airspace. Ethereum smart contracts are used to develop this ecosystem.

- Chain of Things [20]: This is a combined blockchain integrated IoT hardware solution enabled by Maru, a blockchain-IoT integrated research lab. The main role of this application is to provide devices a digital identity when they are first introduced to the market to provide security. There are three major projects in the Chain of Things:
  - Chain of Security: Its main aim is to provide security to IoT through blockchain.
  - Chain of Solar: Functions on connecting solar panels through blockchain to store the produced energy for various applications and the same data can also be calibrated for research purposes.
  - Chain of Shipping: It provides and manages security for shipping, loading, and logistics industry. For monitoring data, automated data logging devices are used which monitor, store, and send the data to the main framework of the network. This application is also developed on Ethereum in terms of the concept of proof.

- Chronicled [21]: This system was created with one goal in mind, to provide the securest ecosystem of IoT with its supply chain. This format has developed several IoT equipment and virtual projects which have cryptographic property. This platform runs on multiple platforms, like Quorum, Ethereum, and Hyperledger, and is done by coordinating multiple blockchain servers at once.

- Modum [22]: This platform's main aim is to improve security and provide data integrity for physical products in order to enhance the supply chain processes. It has been designed to work with multiple blockchain platforms. It stores and senses the environmental conditions during shipments. It has also been used for the distribution of medical equipment and products. The data from the sensors are approved by Ethereum smart contracts. This is how the functions of tagged sensors are used, which is directly linked to the mobile application ownership, which is directly linked to the sensors. At the end an automated recorded ledger analyzes the collected data after the reception of the shipment.

- Twin of Things [23]: This platform ensures the ownership of the IoT objects we use every day. It is developed by riddle and code. The combination of blockchain and cryptography creates a digital identity that is used for hardware devices and all the physical devices in the network. The interaction between these devices includes several transactions that can perform under full secrecy due to the blockchain mechanism. For enabling the device in the blockchain node, a highly secure crypto-chip is provided, which is used in the form of adhesive non-removable NFC tag. An android application is developed for carrying out the blockchain transactions to register the unique and tamper-proof identity of the chip. After proper validation, it becomes part of the node and can interact with other devices in the network.

- Blockchain of Things [24-30]: This is an integrated platform of IoT and blockchain, more simply known as the industrial integrated IoT, which is integrated for the secure and open gateway of communication. It has a Catenis web service layer for rapid blockchain integration combined with an end-to-end encryption. This platform is adaptable in several other platforms of blockchain-like platforms like Ethereum, Hyperledger, etc.

2.6 Conclusion

When modern technologies are integrated they always bring many things to the table, some of which are controversial due to their disadvantages. We have seen the potential of blockchain, which even without a proper governing structure is powerful enough for handling several cryptocurrencies around the world. Blockchain is part of the near future because things are getting digitalized at a very great pace, so there is no doubt that blockchain is here to stay.

Blockchain platforms also have some major issues which can collide with the issues of IoT. Proper consideration should be given when implementing an integration of IoT and blockchain and their various domains. There is no doubt the IoT technology is fragile. There are several other technologies, such as artificial intelligence, which can be used to countermeasure the flaws of IoT. However, the use of blockchain not only provides security to the IoT, it totally revolutionizes it.

Blockchain raises major legal and compliance issues which cannot be ignored, even if we neglect other issues concerning integrated IoT and blockchain. These legal and compliance issues will remain the biggest challenge and will cause many problems in the integration of both IoT and blockchain. Blockchain must have a governing body or admin support that can monitor not just the blockchain technology and their domains, but can also monitor the integration of blockchain with IoT and other technologies. A governing body of blockchain will give it a chance to be used in wide-ranging areas of domains and technologies for many purposes that can be useful in the face-paced growth of modern technologies.

When integrating blockchain with several domains of the IoT for cost-saving and to attain some advantages, the risk of implementation should not be overlooked. The integration of both the technologies should be properly analyzed as to what pros and cons the integration would bring to the modern world and how it will affect the fast-paced growth of the new technologies around us.

In the coming days, blockchain will revolutionize the services and domains of the IoT, and the integration of both technologies should be addressed properly. The integration of both technologies can make the infrastructure of our digital lives easy and secure. After combining these two technologies, we have seen various applications that are smooth and beneficial to mankind in so many ways. IoT has been used in so many different sectors of governments around the world but it comes with several flaws due to its centralized system. The use of blockchain has improved and improvised several IoT services and domains but the integrated use of IoT-blockchain is mainly used for business purposes like smart contracts, secure flow of cryptocurrencies and many more. However, this integration should be used more in the health and defense sectors. The integration of both IoT and blockchain can revolutionize healthcare sectors with IoT's automation and capacity to sense the environment; and the secured features of blockchain can be used in critical domains of the healthcare industry.

The medical supply chain is one of the most important things in today's world and COVID-19 has tested every medical system all around the world. The integration of IoT and blockchain can play a key role in the medical and pharmaceutical industry in the future. Blockchain and IoT can be used to share secure research in the medical field and at a time of crisis around the world, the IoT will provide the data and blockchain will authenticate it, which will help to countermeasure pandemics like COVID-19.

However, for blockchain to function properly, it should be authorized or must have a governing body because this technology doesn't have any legal and compliance requirements. Therefore, various blockchain platforms, like cryptocurrency such as bitcoin, are used by criminals, which raises the issue of trust and blocks this technology from testing its potential.

For all the new technologies, security is a must-have option. The integration of IoT and blockchain has shown how a integration between two new fast-growing technologies can make a big difference. The same idea of implementation should be applied to other new technologies in the form of integration with new technologies, such as artificial intelligence, machine learning, and nanotechnology, but the integration should be done with proper analysis. The integration of new modern technology can be a game changer in terms of the modern development of mankind. The integration of modern technologies might be a possible solution to the many issues we are facing each day.

References

1. Dorri, A., Kanhere, S. S., & Jurdak, R. (2016). Blockchain in internet of things: challenges and solutions. *arXiv preprint arXiv:1608.05187*.

2. Gupta, N., Jha, A., & Roy, P. (2016). *Adopting Blockchain Technology for Electronic Health Record Interoperability*.

3. Reyna, A., Martín, C., Chen, J., Soler, E., & Díaz, M. (2018). On blockchain and its integration with IoT. Challenges and opportunities. *Future Generation Computer Systems*, 88, 173-190. DOI: 10.1016/j.future.2018.05.046.

4. Conoscenti, M., Vetro, A., & De Martin, J. C. (2016, November). Blockchain for the Internet of Things: A systematic literature review. In *2016 IEEE/ACS 13th International Conference of Computer Systems and Applications (AICCSA)* (pp. 1-6). IEEE. DOI: 10.1109/aiccsa.2016.7945805

5. Christidis, K., & Devetsikiotis, M. (2016). Blockchains and smart contracts for the internet of things. *IEEE Access*, 4, 2292-2303.

6. Jentzsch, C. (2016). Decentralized autonomous organization to automate governance. *White paper*, November.

7. Gubbi, J., Buyya, R., Marusic, S., & Palaniswami, M. (2013). Internet of Things (IoT): A vision, architectural elements, and future directions. *Future Generation Computer Systems*, 29(7), 1645-1660. DOI: 10.1016/j.future.2013.01.010.

8. V. Buterin, Ethereum white paper, Available online: https://github.com/ethereum/wiki/wiki/White-Paper, 2013. Accessed: 2018-04-02

9. E. Androulaki, A. Barger, V. Bortnikov, C. Cachin, K. Christidis, A. De Caro, D. Enyeart, C. Ferris, G. Laventman, Y. Manevich, et al. (2018), Hyperledger fabric: A distributed operating system for permissioned blockchains, *arXiv preprint arXiv:1801.10228 (2018).*

10. Litecoin, https://litecoin.org Accessed

11. Quorum whitepaper, Available online: https://github.com/jpmorganchase/quorum-docs/blob/master/ Quorum%20Whitepaper%20v0.1.pdf, 2016.

12. Samaniego, M., Jamsrandorj, U., & Deters, R. (2016, December). Blockchain as a Service for IoT. In *2016 IEEE international conference on internet of things (iThings) and IEEE green computing and communications (GreenCom) and IEEE cyber, physical and social computing (CPSCom) and IEEE smart data (SmartData)* (pp. 433-436). IEEE.

13. Torkaman, A., & Seyyedi, M. A. (2016). Analyzing IoT reference architecture models. *International Journal of Computer Science and Software Engineering*, 5(8), 154.

14. Boudguiga, A., Bouzerna, N., Granboulan, L., Olivereau, A., Quesnel, F., Roger, A., & Sirdey, R. (2017, April). Towards better availability and accountability for iot updates by means of a blockchain. In *2017 IEEE European Symposium on Security and Privacy Workshops (EuroS&PW)* (pp. 50-58). IEEE. DOI: 10.1109/eurospw.2017.50

15. Alenezi, A., Zulkipli, N. H. N., Atlam, H. F., Walters, R. J., & Wills, G. B. (2017, April). The Impact of Cloud Forensic Readiness on Security. In *Proceedings of the 7th International Conference on Cloud Computing and Services Science (CLOSER 2017)*, (pp. 511–517). DOI: 10.5220/0006332705390545

16. Atlam, H. F., Attiya, G., & El-Fishawy, N. (2017). Integration of color and texture features in CBIR system. *International Journal of Computer Applications*, 164(3), 23-29. DOI: 10.5120/ijca2017913600

17. Lo3energy, Available online: https://lo3energy.com/ , 2017.

18. My bit, Available online: https://mybit.io/.

19. AeroToken, Available online: https://aerotoken.org/.

20. Chain of things, Available online: https://www.chainofthings.com/.

21. Chronicled, Available online: https://chronicled.com/.

22. modum, Available online: https://modum.io/.

23. Riddle and code, Available online: https://www.riddleandcode.com

24. Reyna, A., Martín, C., Chen, J., Soler, E., & Díaz, M. (2018). On blockchain and its integration with IoT. Challenges and opportunities. *Future Generation Computer Systems*, 88, 173-190. DOI: 10.1016/j.future.2018.05.046.

25. Atlam, H. F., Alenezi, A., Alassafi, M. O., & Wills, G. B. (2018). Blockchain with Internet of things: Benefits, challenges, and future directions. *International Journal of Intelligent Systems & Applications*, 10(6), 40-48. DOI: 10.5815/ijisa.2018.06.05.

26. Panarello, A., Tapas, N., Merlino, G., Longo, F., & Puliafito, A. (2018). Blockchain and iot integration: A systematic survey. *Sensors*, 18(8), 2575. DOI: 10.3390/s18082575.

27. Shrivastava, G., Le, D. N., & Sharma, K. (Eds.). (2020). *Cryptocurrencies and blockchain technology applications*. John Wiley & Sons. DOI: 10.1002/9781119621201.

28. Kumar, A., Dhanagopal, R., Albreem, M. A., & Le, D. N. (2021). A comprehensive study on the role of advanced technologies in 5G based smart hospital. *Alexandria Engineering Journal*, 60(6), 5527-5536. DOI: 10.1016/j.aej.2021.04.016

29. Gaur, L., Afaq, A., Solanki, A., Singh, G., Sharma, S., Jhanjhi, N. Z., ... & Le, D. N. (2021). Capitalizing on big data and revolutionary 5G technology: Extracting and visualizing ratings and reviews of global chain hotels. *Computers & Electrical Engineering*, 95, 107374. DOI: 10.1016/j.compeleceng.2021.107374.

30. Nartey, C., Tchao, E. T., Gadze, J. D., Keelson, E., Klogo, G. S., Kommey, B., & Diawuo, K. (2021). On blockchain and IoT integration platforms: current implementation challenges and future perspectives. *Wireless Communications and Mobile Computing*. DOI: 10.1155/2021/6672482.

3

Data Communication and Information Exchange in Distributed IoT Environment: Issues and Their Solutions

Rachna Jain[1], Kanta Prasad Sharma[1], Rana Majumdar[3], Dac-Nhuong Le[4]

[1] Amity University Tashkent, Uzbekistan
[2] GLA University Mathura, India
[3] MSIT Techno India Group, India
[4] Faculy of Information Technology, Haiphong University, Haiphong, Vietnam

Email: tokpsharma@gmail.com, nhuongld@dhhp.edu.vn

Abstract

The idea of the internet of things (IoT) has developed over the years in various different phases. It is absolutely true that the world is undergoing consistent transformations that change the trajectory and archives of civilization by some means. This is illustrated by the first and second business revolutions and the information revolution. The IoT is a paradigm based totally on the internet that contains many interconnected technologies like radio frequency identity and Wi-Fi sensor and actor networks to exchange data.

Keywords: IoT, Wi-Fi, wireless network technology, radio frequency

3.1 Introduction

The IoT is a paradigm based totally on the internet that contains many interconnected technologies like radio frequency identity and Wi-Fi sensor and actor networks to exchange data. Agencies, organizations, and small and medium-sized businesses have observed enormous changes in recent years, extending from the regular size of corporations to the universal growth of initiatives and other projects on a mass level.

Globalization plays an essential role in increasing the complexity of project initiatives. The progression of information technology and the IoT is imparting new resolutions, equipment, and is assisting IT programmers and testers in operations of various disciplines in different situations [1,2].

The present requirements for improved mechanisms to monitor and manage many areas, and the continuing research in the field of IoT, have led to the advent and creation of multiple systems like smart home, smart city and smart country. We can find IoT services [3] in two different architectures; one is centralized and the other is distributed. It should be noted that the centralized approach works where the entities acquire and process information centrally, while in the distributed environment, entities exchange data and information at the edge of the network system or topology in a dynamic manner.

There may no longer be a trendy definition of the technologies used in IoT environment; in fact, it depends on the real-time access. Nowadays, the IoT sets the latest trends in accessing verified and reliable information and data analysis statistics [18-33]. In fact, all scientists and researchers have come up with some meaningful definitions of the same technology, as well as all the business persons and company owners who have provided extraordinary capabilities and particularities.

Madakam *et al.* [3] have given the possible definition of IoT as "An open and comprehensive network of intelligent objects that have the capacity to auto-organize, share information, data and resources, reacting and acting in the face of situations and changes in the environment."

This chapter will not only help you recognize the dissimilar challenges [19, 20] and security threats in the internet of things environment and cloud system but also help you increase your knowledge about different interesting characteristics and strengths. In this chapter we present the main idea behind data communication and information exchange [33] issues and challenges in IoT environment with their possible solutions.

The Beginning of IoT

The IoT was undoubtedly discussed and described by Ashton in 1999 [3]. One can understand IoT as an interconnected set of different things like humans, data, information, tags, etc., which is found, processed and available over the internet and able to communicate and respond all over the world. The basic idea behind IoT is to get meaningful data or information about the environment in order to know, control, understand, and act on it [2].

At present, the IoT concept is being explored like anything else. It covers various different technologies, standards and services which communicate with this digital world and process information over the internet [3,4]. A typical IoT environment consists of various smart devices and systems that interact with each other over the internet to process information on the same or different platforms to fulfill a common goal. The IoT platforms deploy various different communication architecture and processing units and topologies based on their final requirements. For example, the IoT system explores the properties and controls the capabilities of wireless networks. It also plays a vital role in data communication and exchange in centralized as well as distributed environments [3].

As previously mentioned, the World Wide Web (WWW) has shifted from its conservative perception (that allows client server technology and serves as network connections with various different users) to covering the methods for connecting the physical world to the virtual one. Researchers refer to this concept and perspective as the Internet of Things (IoT) [12].

Because the IoT is established on the perception of smart objects or smart things, it is important to deliver a perfect description of "smart things," sometimes called "smart objects." For instance, according to Miorandi, smart objects are accessible as physical things that are acknowledged by specific features that rely on an identifier [12].

Their ability to be recognized and reply to any incoming signals is negligible and can be identified by the address and name. Smart objects have the capability to both compute and detect physical things like heat or light, and also simulate action.

The various technologies related to the IoT environment are shown in Figure 3.1.

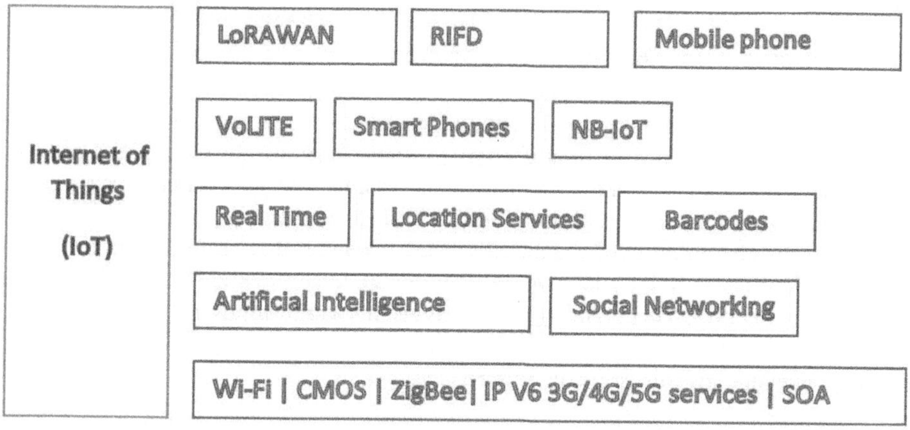

Figure 3.1: Various technologies related to the IoT environment.

3.2 IoT Technologies and Their Uses

3.2.1 How WSN Works

Wireless sensor networks (WSNs) [14] or Wi-Fi sensor networks take a brand new hypothesis of real-time entrenched structure with other prospective systems in our everyday existence wherein conventional substructure totally based on network is almost infeasible. The sensor device [28,31] includes a transceiver, a tiny implanted processor, current supply, interface unit and a space storage unit used for collecting information of the real-time processing data and send it to the data access controllers [6,8].

3.2.2 Communication with RIFD-Enabled Devices

Radio frequency identification (RIFD) is a form of wireless communication [4,5] that incorporates the use of electromagnetic or electrostatic coupling with radio frequency (RF), which is basically used for identifying the objects, smart things, living things, etc. [8,11].

3.2.3 WWW - Things on the Web

Smart objects are part of World Wide Web (W3C), which is also an essential part of the Internet of Things environment. Nowadays, the internet utilizes the concept of "Web of Things" and Web 2.0 technologies, which make use of various scripting languages like Java and Ajax XML. For example, Ajax plays a vital role in Web 2.0 [9] and is used to overcome the latency between the web server and client [29]. Unlike Web 1.0, it is a simplex communication (one-way communication system) and only provides read only data on the web, whereas Web 2.0 works in full duplex mode (two-way communication system) and provides both read only and writable content on the internet [10].

Currently underway, Web 3.0 is likely to be introduced soon. It is a concept where all the data and information can be accessed or stored in a systematic manner and this information can be understood by both the human and the machine. A leading example of Web 3.0 is a virtual shopping mall.

3.3 Centralized vs. Distributed Approach

These days, the international cloud continues to be extended and has taken many forms. From the virtual partitions on mainframes to virtualization, cloud services and cellular technology, the cloud is composed of diverse systems and approaches. We begin with centralized systems as they are thoroughgoing and easy to describe [7].

Centralized architectures are generally those systems that use the traditional client-server technology in which one or more than one end user or node are at once connected to a significant server. This is the most usually used form of topology in most of the organizations and companies where the client node sends the data request to the server and in turn it receives the information (see Figure 3.2).

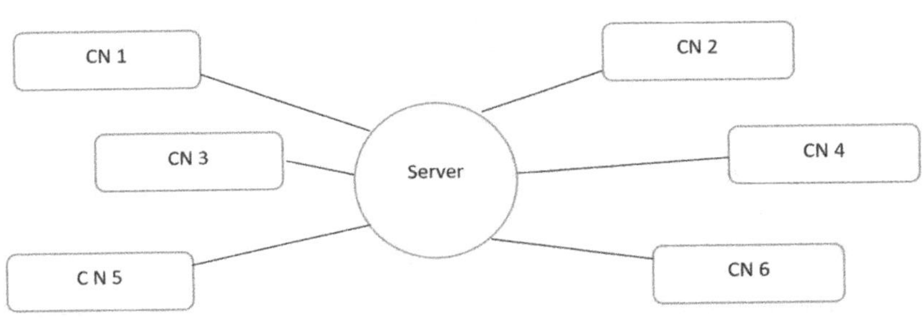

Figure 3.2: A centralized system in IoT environment (CN = Client Node).

Wikipedia, a free online encyclopedia, is a best fit example in this case. Remember that when we request that a server (which has big data to analyze) retrieve the desired information which the client node is looking for, we actually need a centralized information server. Suppose someone is really scared and wants to fetch the full information on the COVID-19 pandemic and its causes. Then, in a simple way they may send the request to the server of Wikipedia (located in Virginia, USA), which then replies with the searched articles and information based on the search criteria by the client node. Now, one can get the idea that the Wiki database system is a Server Node and the use is Client Node.

3.3.1 Centralized System and Its Physiognomies

We can characterize a centralized system as having a few important characteristics, which are very fundamental and easy to understand.

- Global clock: As we know, the architecture of a centralized system consists of a node called a server or master node and various interconnected clients nodes, generally called a slave or end user. These client nodes are synchronized with the help of a global clock, sometimes called a central clock.
- Central unit: Each centralized system should have at least one central unit which is used to control all other units or nodes in the system, which are interconnected with the server.
- Server failure dependency: This is one of the important characteristics of a centralized environment where a client's nodes have dependency on server node. Once a server node has fallen down or failed, no one can communicate with the other because central or server failure causes the entire system to fail.
- Scaling in centralized system: Typically two types of scaling are possible; one is vertical scaling and the other is horizontal scaling. In centralized mode only vertical scaling is possible at server node. However, there is a limit to scaling up vertically.

3.3.2 Advantages

Physically secure: Client server architecture in centralized node offers ease of physical security. Due to its location the client node is very much unyielding in the system, which makes it easy to provide security to the server.

Allows use of dedicated resource: One of the main advantage of a centralized system is dedicated resources for each client, e.g., HD, memories, virtual space, etc.

Cost effective: At a certain limit it seems to be very cost-effective for small business organizations, as a centralized system has only one server, which is again a cost-effective system, since more servers are more expensive.

Random update is very easy: When a random update needs to be done, it's an easy task because there is only one system that needs to be updated, which is only the server. No client machine needs to be updated.

Detachment of client node is very easy: In centralized system technology the detachment of any client node is very easy, all that needs to be done is to disconnect it from the server.

3.3.3 Disadvantages of Centralized System

Network connectivity is essential: The centralized system has only one server or central node for data fetching and processing. In this scenario, network connectivity is very essential.

Data backup: Data backup in centralized node is very risky because once the server node fails than one can lose the data immediately.

3.4 Distributed System Architecture

In distributed systems [25], each user makes their own selection. The performance of the system is the cumulative selections of the discrete system and nodes. The idea of a distributed IoT [26] has a wide range of diversity. Don't forget that it is one of the feasible techniques which could push the nightmare of the IoT into the physical world. In the last few years, scientists and researchers have come forward with various studies in the field of distributed IoT architecture and environment. Let us take an example of Ning and Liu [30], which provides a new horizon and views based on a new hybrid system called U2IoT that consists of two subsystems. One is called Unit IoT, and the other is known as Ubiquitous IoT, which includes various other different Unit IoTs (see Figure 3.3).

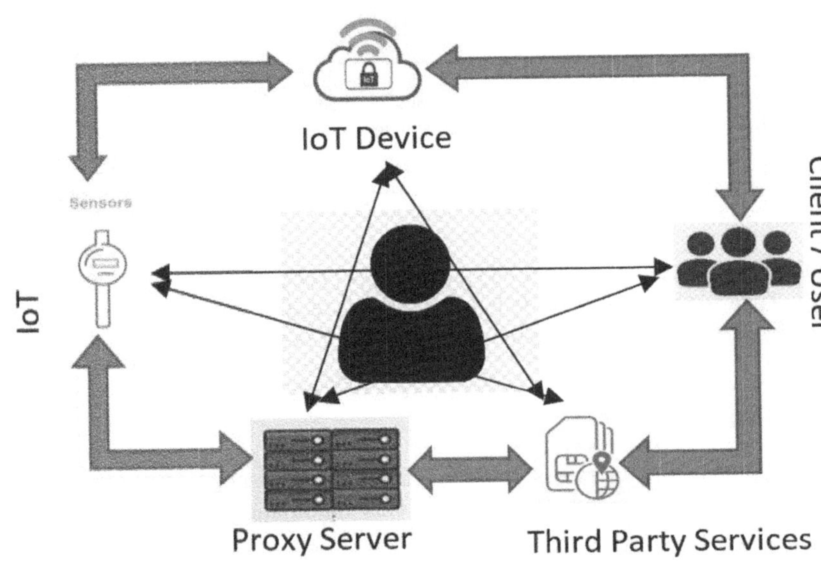

Figure 3.3: Distributed architecture in IoT environment.

For instance, metadata is data that contains data about other data, which is exactly the key feature of the Google search engine system. When a search request is generated by the client on Google server, hundreds of computer systems work on it and generate every possible data about the data which is actually being searched for by the client node. For the client, it seems to appear on his/her single system, but in actuality, multiple computer systems work together on this single request, which is returned as searched query.

3.4.1 Advantages of Distributed System Architecture

Physically secure: Client server architecture in distributed mode offers ease of physical security. Because each client has their own server node, it is very much unyielding in the system; thus, it is easy to provide security to the server [26].

No global/central lock: In distributed mode each user or node is an autonomous node and does not depend on other nodes, and therefore they have different autonomous clocks that they run and track.

Peer-to-peer: One of the best advantages of a distributed system is that all nodes participate with each other and work towards a common goal.

More than one server/central unit: In this mode every system is treated as a server; therefore, in case of system failure, we have the chance to access data or information from any other server or node which is available to access and listen to the commands.

Scaling in distributed system: Typically, two types of scaling are possible; one is vertical scaling and the other is horizontal scaling. In distributed mode, both types of scaling are possible at server node. There is no limit to scale up vertically or horizontally [25].

Low latency compared to centralized system: We all know that the distributed system environment has low latency because of its very high geographical range. Therefore, it takes much less time to get a response compared to centralized systems, which take a little more time to get a response.

3.4.2 Drawbacks of Distributed System Architecture

Challenging to detect failure: The distributed system has multiple servers or central nodes for data fetching and processing. In this scenario, if a failure occurs in network connectivity it is very difficult to determine, and therefore rectify, the node that has actually failed.

Security challenges in distributed systems: Though conventional studies on safety in the IoT environment are still at the beginning stage, there should be a significant body that analyzes the prevailing challenges and feasible protection methodologies. However, in the current research, publications are mainly used to provide an introductory part of generic issues without the consideration of other issues like data communication and information exchange in distributed environment [25, 26]. To know about the specific security issues of a distributed system that works in an IoT environment one should know about the necessity of analyzing the risks associated with distributed IoT principals over various safety threats.

In order to evaluate the associated risks and security threats, we need to know about the significant experiments in the strategy and implementation of safety mechanism. These studies also point out the specific issues related to data communication and information exchange in distributed IoT environment. As we know, in the world of IoT security the major challenge for researchers and scientists is how to restrict the successful deployment of distributed principals in an IoT world.

It is now a known fact that IoT architecture reacts with a projected population of millions of smart things on the WWW, which defiantly interact with each other and are called artifacts [8]. Now all these communications must somehow be protected. However, this task is a very complicated and difficult one. Examples of some previously explained security challenges and threats are explained below:

- Smart connectivity: IoT devices and sensors that are related and communicated with each other through the WWW and IoT environment might also want to update their developments, which in turn should be projected onto the adjustments of neighboring environments. The IoT [1, 2] is a new dynamic and intelligent infrastructure that can analyze the calculated data and make the desired selections to enhance and change itself by changing features of the connected devices to accommodate the surrounding environments' amendments. The IoT era is an intelligent and smart era that facilitates all smart devices [1, 2] that are connected with each other to update themselves and are consistent with modifications within the neighboring environment [9]. As a result, smart objects and intelligent devices can be manufactured if smart infrastructure is nicely designed to deal with the collected and processed information from various smart objects or devices efficiently, to make the required decisions.

- High privacy and security: The key idea behind the use of the internet of things is to have smart devices that communicate with millions or billions of devices over the entire real world. To achieve a high level of security and protection is a very big task for IoT [7].

3.5 Data Communication Taking Place in Distributed IoT Environment

3.5.1 Internet of Things (IoT) Protocol

The technical communication between smart objects, Wi-Fi sensors, wireless devices, servers and many other user applications, runs with the help of a new platform called IoT platform. All of these are also the necessary components of IoT. To enable communication between smart objects, sensors, Wi-Fi-enabled devices and networks actually requires a protocol suite. These protocols are especially designed and developed for IoT-like environments. Three of the following protocols are useful for data communication during IoT distributed enlivenment.

3.5.2 Constrained Application Protocol (CoAP)

This protocol is used for the HTTP model and was designed to translate HTTP in restrictive device and network environments. The constrained application protocol (CoAP) [15,17] depends on the user datagram protocol (UDP) for making communication between the end points secure. The main advantage of UDP is that it transmits data to multiple hosts while using retaining communication and minimum bandwidth. However, the architecture of HTTP supports communications between applications in the form of request and response. The main objective of CoAP is controlling the message services and marking all messages as "confirmable" and "non-confirmable" [11].

3.5.3 Message Queuing Telemetry Transport (MQTT)

The message queuing telemetry transport (MQTT) protocol is the most effectively used and adopted protocol suite in the IoT environment. Telemetry transport (TT) is a very lightweight type messaging protocol. This protocol is especially designed for smart devices which are battery operated. This protocol works on top of TCP/IP and is especially designed for unreliable data transfer. Message queuing telemetry transport is based on subscriber and publisher model, where the publisher is responsible for collecting the data and sending this information to all subscribers with the help of a mediation layer (see Figure 3.4).

The MQTT protocol provides three types of services in this mode, which is called quality of service (QoS):

- At most once: This is one of the fastest modes without confirmation.
- At least once: This service ensures that the message is diverted at least once. But this service is not restricted on duplicates.
- Exactly once: This is the most reliable service.

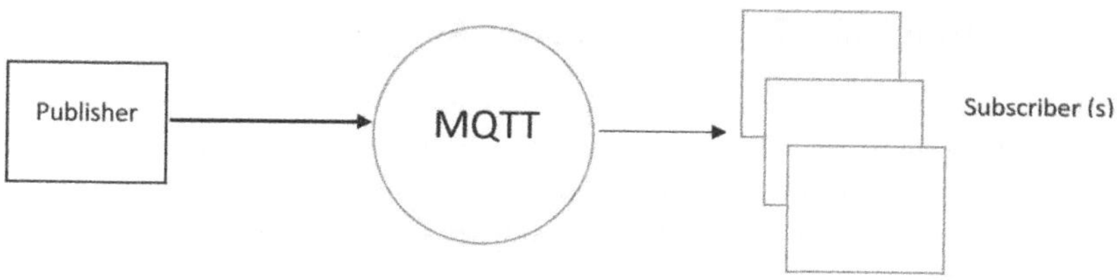

Figure 3.4: The MQTT protocol.

3.5.4 Wi-Fi

A Wi-Fi communication [15, 25, 30] held between two smart devices is able to transmit wireless signals from computer systems, routers, hubs, mobile phones, etc. This protocol provides internet connectivity [32] in terms of public or private connection within a home or office or within a wide range of areas. These devices which connect with Wi-Fi connection can connect within a certain range. A Wi-Fi hotspot is another way to connect with nearest Wi-Fi devices. In an IoT system, Wi-Fi uses radio frequency to broadcast information at some particular frequencies of up to 5 GHz channels.

3.5.5 Zigbee

The specialization of Zigbee-based communication [27] networks is minimum power consumption and minimum throughputs which ranges up to 250 kbps at the 100 meters between nodes. Zigbee communication networks include applications like sensor networks, personal networks, etc. They are easy to install and maintain. They implement healing grid topology and easily scale a huge number of nodes.

3.5.6 Extensible Messaging and Presence Protocol (XMPP)

This protocol was developed by Jabber open source community in 1999. It was developed for real-time messaging services and communication. This is another IoT communication protocol based on XML languages that allows real-time exchange of messages between two or more subscribers.

3.5.7 Data Distribution Service (DDS)

This is another protocol that deals with IoT. Originally designed by Object Management Group, it was initially based on publish-subscribe technology. For getting reliable, high-performance, real-time M2M communication, DDS protocol-enabled technology is used in IoT devices to manage big data [13]. The architecture of data distribution services protocol is supported by a data-centric pub-sub (DCPS) layer and data local reconstruction layer (DLRL).

3.5.8 Advanced Message Queuing Protocol (AMQP)

This is an open standard protocol designed in 2003 specially dedicated to financial services. This protocol is used in messaging features, queuing, routing and security specification. The most technological use of AMQP is in robust communication prototypes.

3.5.9 Smart Home and IoT Applications: An Example

Now, if someone is considering playing with IoT environment and technology they, must interact with IoT applications. When we think about IoT application, the first thing that comes to mind is "Smart Homes" and the unique example of this is known as "JARVIS," which is a truly automated system based on artificial intelligence that is owned by the famous Mark Zuckerberg (see Figure 3.5).

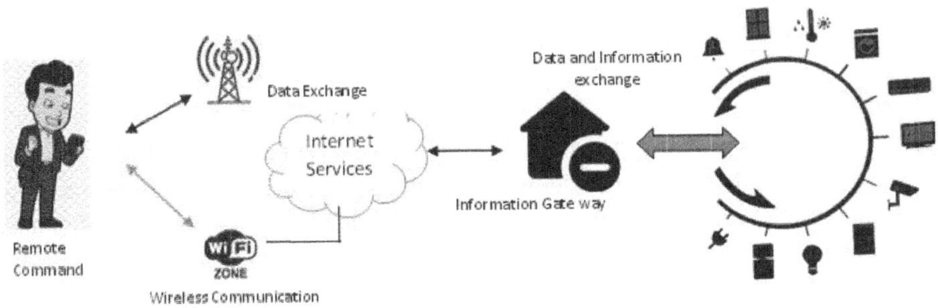

Figure 3.5: How a smart home works in IoT environment.

3.5.10 IoT Services, Machines and Applications

The IoT has found applications in several areas like manufacturing, healthcare, transportation, and smart cities and homes. This review focuses on manufacturing execution systems (MES) and sensor-based modeling of manufacturing systems, and therefore the recent development and application of IoT technologies within the manufacturing domain.

3.5.11 Sensor-Based IoT Services

Advanced sensing ends up as the large amount of information inhabiting ERP, MES, and PCS. Currently, a big quantity of information already exists within the manufacturing domain; however, it isn't absolutely used for period method watching, fault designation, and performance improvement. Realizing the total potential of MES and advanced sensing depends on the event of recent methodologies to extract helpful options and patterns from the information, then exploit the new data to alter good production. Here, we tend to categorize sensor-based manufacturing information processing and management into four specific areas as follows:

- Visualization of data
- Feature extraction and image pattern recognition
- Sensor data fusion

3.5.12 Application in IoT Environment

Two basic applications have received increasing attention over the past 10 years. Above all, industrial IoT has yielded the quickest increase over the past three years. The IoT provides the following applications:

- Cloud-based applications on IoT platform.

- Application systems related to cyber security.

3.5.13 Future of IoT in a COVID-19 Pandemic

In the event that COVID-19 spreads throughout the entire world again, artificial intelligence (AI) might help businesses efficiently plan for supplies while also assisting in the prognostication of consumers' desires, which have become practically predictable. Chatbots may provide customers with help around-the-clock, making them a "must-have" during sequestration. The need for the creation of algorithm-based moderators of posts and visual content on social networks may help machine learning gain more acceptance.

Internet of things (IoT) gadgets change how computers, cellphones, and other devices are connected to the internet. They have a variety of sensors, built-in technology, and deliberate programming put into them [31-33].

IoT devices are primarily components of smart cities, smart homes, wearable technology, safety monitoring, and waste management. With the help of this technology, India may be able to identify and cure people's health issues even before any symptoms manifest. Furthermore, it may be possible to use much more personalized techniques for prescribing medications and administering therapies. There were close to 30 billion IoT devices in the middle of 2019 and 2020 [24], and statista.com predicts that number would rise to 30.73 billion by the end of 2020 and to 75.44 billion by 2025 [34,35].

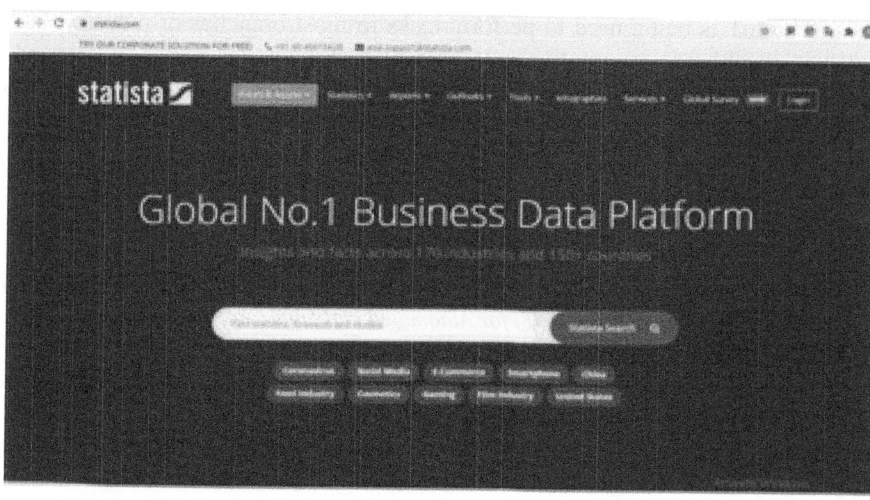

Figure 3.6: A global business data platform. (Source: www.statista.com)

Some of the major dangers to this continuing trend are lack of security, absence of universal international compatibility, and a manageable decrease in manual chores. At the same time, the IoT provides sensible management and automation, saves cash and time, and will offer a much better quality of life (see Figure 3.6).

3.6 Conclusion

The data communication and data exchange in IoT environment presents various advantages to customers and has the potential to alter the basic ways in which customers use technology. In the future, the IoT will probably meld the virtual and physical worlds in ways which are presently hard to grasp.

From a security and privacy perspective, the expected pervasive introduction of sensors and devices into presently intimate areas, such as homes, cars, and even the body with wearables and ingestibles, pose explicit challenges. Even though we continue to share observations regarding aspects of our lives more and more, we seemingly still need privacy.

Employees can still enforce laws, educate customers and businesses, and have interaction with client advocates, industry, academics, and different stakeholders concerned within the IoT to market acceptable security and privacy protections. At the same time, we have a tendency to urge any self-regulatory efforts on the IoT, besides the enactment of knowledge security and broad-based privacy legislation.

There is plenty of analysis on many alternative areas involving the IoT. Many researchers have proposed many alternative sorts of variations of protocols and ways to authenticate the IoT, which makes it terribly difficult to spot the most effective resolution. Therefore, there are structured tips within the kind of standardization so as to interconnect every kind of device, protocol, application, etc. Developing standards or solutions has to be associated with open supply protocols and ways so as to draw in wide acceptance and use. We hope that by attempting to explain how such a standard should be produced and what requirements are necessary, we have laid the groundwork for future research in the area.

The IoT has continued to expand over the last three decades. Data communication in IoT distributed environment is always a typical task when transferring information within smart devices. The IoT is being used to perform tasks in most branches of industry, including the financial, healthcare, automotive and transportation industries, and is still expanding its range. This is the reason why data communication is a critical task in the IoT environment and why data exchange is now necessary for most reliable and effective communication.

References

1. Ismail, Y. (2019). Introductory Chapter: Internet of Things (IoT) Importance and Its Applications. In *Internet of Things (IoT) for Automated and Smart Applications*. IntechOpen. DOI: 10.5772/intechopen.90022, 2019
2. Kumar, S. (2019). A Review on client-server based applications and research opportunity. *International Journal of Recent Scientific Research*, 10(7), 33857-3386.
3. Madakam, S., Lake, V., Lake, V., & Lake, V. (2015). Internet of Things (IoT): A literature review. *Journal of Computer and Communications*, 3(05), 164.
4. Brandt, A., & Buron, J. (2014). Transmission of IPv6 Packets over ITU-T G.9959 Networks. *RFC 7428* (Proposed Standard).
5. Elfrink, W. (2014). The internet of things: Capturing the accelerated opportunity. *Cisco Blog*, 15.
6. Fielding, R., & Reschke, J. (2013). Hypertext Transfer Protocol (HTTP/1.1): Message Syntax and Routing. *RFC 7230 (Proposed Standard)*.
7. Cirani, S., Picone, M., Gonizzi, P., Veltri, L., & Ferrari, G. (2014). IoT-OAS: An oauth-based authorization service architecture for secure services in iot scenarios. *IEEE Sensors Journal*, 15(2), 1224-1234. DOI: 10.1109/jsen.2014.2361406.
8. Teklemariam, G. K., Hoebeke, J., Van den Abeele, F., Moerman, I., & Demeester, P. (2014, September). Simple RESTful sensor application development model using CoAP. In *39th Annual IEEE Conference on Local Computer Networks Workshops* (pp. 552-556). IEEE. DOI: 10.1109/lcnw.2014.6927702.

9. Silverajan, B., & Savolainen, T. (2014). CoAP Communication with Alternative Transports. *Draft-silverajan-core-coapalternative-transports-06 (work in progress).*
10. Gartner (2013). Gartner Says the Internet of Things Installed Base Will Grow to 26 Billion Units By 2020. *Gartner.*
11. Shelby, Z., Hartke, K., & Bormann, C. (2014). The constrained application protocol (CoAP). *RFC 7252.*
12. Waller, M., & Fawcett, S. (2013). Data Science, Predictive Analytics, and Big Data: A Revolution That Will Transform Supply Chain Design and Management. *Journal of Business Logistics*, 36(2), 77-84. DOI: 10.1111/jbl.12010.
13. Evangelatos, O., Samarasinghe, K., & Rolim, J. (2013, May). Syndesi: A framework for creating personalized smart environments using wireless sensor networks. In *2013 IEEE international conference on distributed computing in sensor systems* (pp. 325-330). IEEE. DOI: 10.1109/dcoss.2013.35.
14. Teklemariam, G. K., Hoebeke, J., Moerman, I., & Demeester, P. (2013). Facilitating the creation of IoT applications through conditional observations in CoAP. *EURASIP Journal on Wireless Communications and Networking*, 2013(1), 1-19. DOI: 10.1186/1687-1499-2013-177.
15. Gubbi, J., Buyya, R., Marusic, S., & Palaniswami, M. (2013). Internet of Things (IoT): A vision, architectural elements, and future directions. *Future Generation Computer Systems*, 29(7), 1645-1660. DOI: 10.1016/j.future.2013.01.010.
16. Hersent, O., Boswarthick, D., & Elloumi, O. (2011). *The internet of things: Key applications and protocols*. John Wiley & Sons.
17. Blackstock, M., & Lea, R. (2012, October). IoT mashups with the WoTKit. In *2012 3rd IEEE International Conference on the Internet of Things* (pp. 159-166). IEEE. DOI: 10.1109/iot.2012.6402318.
18. Miorandi, D., Sicari, S., De Pellegrini, F., & Chlamtac, I. (2012). Internet of things: Vision, applications and research challenges. *Ad hoc Networks*, 10(7), 1497-1516. DOI: 10.10.1016/j.adhoc.2012.02.016., 2012.
19. Boswarthick, D., Elloumi, O., & Hersent, O. (Eds.). (2012). *M2M communications: a systems approach*. John Wiley & Sons. DOI: 10.1002/9781119974031.
20. Wilson, D. W., Lin, X., Longstreet, P., & Sarker, S. (2011). Web 2.0: A definition, literature review, and directions for future research. *AIS Electronic Library (AISeL).*
21. Solanki, M. S., Sharma, K. P., Goswami, L., Sikka, R., & Anand, V. (2020, March). Automatic Identification of Temples in Digital Images through Scale Invariant Feature Transform. In *2020 International Conference on Computer Science, Engineering and Applications (ICCSEA)* (pp. 1-6). IEEE. DOI: 10.1109/iccsea49143.2020.9132897.
22. Emmerson, B. (2010). M2M: the Internet of 50 billion devices. *WinWin Magazine*, 1, 19-22.
23. Shelby, Z., & Bormann, C. (2011). 6LoWPAN: The wireless embedded Internet. John Wiley & Sons. (Wiley Series on Communications Networking & Distributed Systems). Wiley, 2010. DOI:10.1002/9780470686218.
24. Beltran, V., Torres, J., & Ayguadé, E. (2008, April). Understanding tuning complexity in multithreaded and hybrid web servers. In *2008 IEEE International Symposium on Parallel and Distributed Processing* (pp. 1-12). IEEE. DOI: 10.1109/ipdps.2008.4536267.
25. Kumara, S. R. T., & Bukkapatnam, S. T. S. (2007). Characterization and monitoring of nonlinear dynamics and chaos in manufacturing enterprise systems. In *Network Science, Nonlinear Science and Infrastructure Systems* (pp. 99-122). Springer, Boston, MA. DOI: 10.1007/0-387-71134-1_5.

26. Pandey, B., & Sharma, K. P. (2019, February). Radar transmogrification technology: Support for unmanned system. In *2019 Amity International Conference on Artificial Intelligence (AICAI)* (pp. 913-917). IEEE. DOI: 10.1109/aicai.2019.8701369.

27. Langendoen, K., Baggio, A., & Visser, O. (2006, April). Murphy loves potatoes: Experiences from a pilot sensor network deployment in precision agriculture. In *Proceedings 20th IEEE International Parallel & Distributed Processing Symposium* (pp. 1-8). IEEE. DOI: 10.1109/ipdps.2006.1639412.

28. Carrera, D., Beltran, V., Torres, J., & Ayguadé, E. (2005, July). A hybrid web server architecture for e-commerce applications. In *11th International Conference on Parallel and Distributed Systems (ICPADS'05)* (Vol. 1, pp. 182-188). IEEE. DOI: 10.1109/icpads.2005.30.

29. Du, W., Deng, J., Han, Y. S., Varshney, P. K., Katz, J., & Khalili, A. (2005). A pairwise key predistribution scheme for wireless sensor networks. *ACM Transactions on Information and System Security (TISSEC)*, 8(2), 228-258. DOI: 10.1145/1065545.1065548.

30. Liu, D., Ning, P., & Li, R. (2005). Establishing pairwise keys in distributed sensor networks. *ACM Transactions on Information and System Security (TISSEC)*, 8(1), 41-77. DOI: 10.1145/1053283.1053287.

31. Holland, G., & Vaidya, N. (2002). Analysis of TCP performance over mobile ad hoc networks. *Wireless Networks*, 8(2), 275-288.

32. Heinzelman, W. R., Chandrakasan, A., & Balakrishnan, H. (2000, January). Energy-efficient communication protocol for wireless microsensor networks. In *Proceedings of the 33rd annual Hawaii international conference on system sciences* (pp. 10-pp). IEEE. DOI: 10.1109/hicss.2000.926982.

33. Sharma, K. P., Poonia, R. C., & Sunda, S. (2018, August). Accurate Real-Time Location Map Matching Algorithm for Large Scale Trajectory Data. In *2018 7th International Conference on Reliability, Infocom Technologies and Optimization (Trends and Future Directions) (ICRITO)* (pp. 646-651). IEEE. DOI: 10.1109/icrito.2018.87487450

34. Puar, V. H., Bhatt, C. M., Hoang, D. M., & Le, D. N. (2018). Communication in internet of things. In *Information Systems Design and Intelligent Applications* (pp. 272-281). Springer, Singapore. DOI: 10.1007/978-981-10-7512-4_28

35. Singh, D. K., Sobti, R., Jain, A., Malik, P. K., & Le, D. N. (2022). LoRa based intelligent soil and weather condition monitoring with internet of things for precision agriculture in smart cities. *IET Communications*, 16(5), 604-618. DOI: 10.1049/cmu2.12352.

4

Contribution of Cloud-Based Services in Post-Pandemic Technology Sustainability and Challenges: A Future Direction

NEERAJ KUMAR PANDEY[1], SAMPOORNA KASHYAP[1], ASHISH SHARMA[2], MANOJ DIWAKAR[3]

[1] DIT University, Dehradun, Uttarakhand, India
[2] GLA University, Mathura, Uttar Pradesh, India
[3] Graphic Era Deemed to be University, Dehradun, Uttarakhand, India
Email: dr.neerajkpandey@gmail.com, skkomalkashyap@gmail.com, ashishs.sharma@gla.ac.in, manoj.diwakar@gmail.com

Abstract

In view of the COVID-19 outbreak, many companies and education systems have enabled their own online mode of working and making their resources available online on their local servers. There is a need to move the entire information and communication technologies (ICT) infrastructure to cloud with the same capability and performance. In a pandemic, users are exposed to various risks due to unsafe network or using public network for confidential tasks. So, the cloud computing environment plays a very significant role in the security and service availability. In this chapter, the contribution of cloud in all major industrial domains is the focus. Specific tools and services available on cloud for addressing dedicated industrial issues are helping in business automation in the era of Industry 4.0. So, with the help of cloud technologies the other backbones, like data analytics, IoT and machine learning, are creating miracles for industries and businesses. The contribution of cloud technologies in agriculture, weather forecasting, medical image analysis, security, ICT, and entertainment is discussed in this chapter with future application and utilities. This chapter also exposes the various applications and tools used in different industrial areas supported by cloud computing.

Keywords: Cloud computing, industry 4.0, security in cloud, IoT, medical imaging

4.1 Introduction

Cloud computing encompasses remote servers, that host software and/or virtual infrastructure and provide them as services to clients over the internet, which can then be accessed and controlled using the web or an API. This model of computing originated in the late 20th century with mainframe computers and early VM operating systems, but the term cloud computing, in today's context, has been around for about a decade. While referring to cloud, one tends to speak about the public cloud. Public cloud generally provides daily used services like email, storage, etc., which can be accessed using the web or a specific application. These services are mostly free and sometimes include infrastructure as well. Private cloud is usually owned by larger corporations and companies for in-house networking and communication; thus, does not share its resources with other tenants. A large chunk of the management of the provided services is done by their employees. A private cloud can also be remotely managed by some cloud service provider, but the resources are not shared with anyone. With the advent of technology and the growth in the demand for cloud-based services, public cloud was not very trustworthy amongst the clients for sharing sensitive data and private cloud lacked versatility. This saw the emergence of the hybrid cloud. Hybrid cloud uses parts of both public and private cloud to provide infrastructure, platform, and software as services. Many mega giants across several sectors use hybrid cloud to provide services which range from entertainment to security.

The coronavirus (COVID-19) pandemic adversely affected the entire world in the spring of 2020. It brought the daily operations of industries across the globe to a standstill. Due to governments urging people to stay indoors, there was a surge in demand for services provided over the internet. From the delivery of medicines and food to educational institutes and multinational companies, web-based services were demanded across all sectors. Cloud-based services saw a major spike which eventually changed how this technology was viewed and used. A considerable amount of data was generated, and these service providers provided a speedy service to quickly tackle it [1]. Major growth and evolution of the Fourth Industrial Revolution (Industry 4.0) was seen during this time. It saw a major development in manufacturing technologies and its gradual shift to automation and data exchange over the internet. Cloud computing along with other emerging technologies like IoT, AI and HCI paved the way for this. Factories under this revolutionary phase utilized machinery supported by remote connections like Wi-Fi and other sensory devices. They further formed an essential part of the network which is connected to remote workstations that can observe, predict, plan, and control the entire production process and its progress. Hence, these industries used smart manufacturing processes for making essential goods during and after the pandemic, like disposable medical equipment and other items to mitigate the crisis in the healthcare sector [2, 3].

The post-pandemic era saw a rise in the intervention of technological and other engineering methods in a wide arena. Cloud services as solutions spread from the IT industry to almost every sector. It saw several industries migrating from traditional methods and networks to cloud technologies and adopting cloud services: either as clients or providers. Be it for educational purposes, research or governance and administration, cloud technology gradually seeped into our everyday lives. With various service models, cloud aided in providing solutions to a multitude of customers to address various issues, globally. Though it is insurmountable to mention the outreach of cloud-based solutions, these few sectors saw a steep rise in demand for the aforementioned.

Agriculture has long evolved from the days of farming crops and rearing livestock for personal consumption and/or for small monetary gain. In fact, the agriculture sector is a major income source for many nations across the world. Technological advancement and scientific research have aided in the development of the agricultural industry. Rural development both economically and socially is extremely dependent on the outreach of the agricultural division. Along with cultivation of crops and rearing of livestock, agriculture also includes irrigation, pre- and post-processing, biotechnology, and other environmental impacts. Cloud-based solutions to aid this sector include cloud-based software for farm management to remotely control the various processes included in the long line of growing crops by an administrator [4]. The quantity of data for storing and processing and learning from them to take correct decisions regarding farming of crops has paved the way for researchers to develop new solutions based on designing cloud-based software services [5]. Cloud storage in this arena took a major spot as old methods of storing these IoT-related data were unfruitful. Cloud-based big data analytics (BDA) and IoT thus have an important role in the further R&D space regarding smart agriculture [6].

Public health and healthcare have been a major concern of governments for several decades. Though healthcare is made accessible to the entire population of a country, public health is a major concern. While healthcare envelopes the overall health and other medical status of a person, public health ensures clean drinking water, proper sanitation, and other hygiene facilities for the populace. The pandemic brought about a major revolution in the use of cloud-based services in these fields. While patient databases were created to track their medical history and health status, departments of public health surveyed water sources, constructed sanitation facilities and also educated people about its importance. Though several technologies were used to attain the aforementioned, cloud-based services were used extensively. Be it for databases, online resources, technical solutions, meetings or even research methods, these services came a long way to aid these departments to thrive in the wake of this unforeseeable circumstance. All necessary information was proposed to be stored and made available from where users could access it via the web, helping in making quick, real-time decisions regarding various diseases and treatments. This technology was suggested to be incorporated with blockchain, etc., to fulfill the needs of ICUs at various levels [7]. Cloud IoT in this sector can further bring about major changes, drastically improvising and leading to its continuous and systematic development [8]. For instance, the Salesforce Care solution was introduced specifically for the frontline workers to continue to fight the COVID-19 virus [9]. Cloud IoT can also be used to get patient information via sensors or other medical devices, then send it to the cloud for storing, processing, and managing information and keeping it as HER for specific usage [8,10,11].

Cloud computing has uplifted information technology to higher limits by offering businesses the background for storing data of flexible capacity which can further be scaled to match their elastic demand and supply, further decreasing their capital expenditure. However, it also comes at the cost of efficiently managing its security [12]. Cloud computing has unique attributes that require risk assessment in areas such as data integrity, recovery, and privacy, and an evaluation of legal issues in areas such as e-discovery, regulatory compliance, and auditing [13]. Security worries begin as apps start running outside of the specified firewall and inch towards the public domain. A huge amount of data is generated around the globe each day. Data security and privacy are major concerns. Though the service providers ensure the security of the clients' data, people mostly prefer storing their essential and important data in their personal workstation. Clients prefer private cloud over public cloud for security concerns as well. A lot of literature has been published to overcome security breaches and cloud service providers are working on the enhancement of this feature. Increasingly, letting a third-party auditor authenticate the reliability of the information stored in the cloud for the client is being practiced [14]. To get effective data dynamics, a better Proof of Retrievability model [14] by modifying the classic Merkle Hash Tree (MHT) construction for block tag authentication has also been proposed [15].

Industry 4.0 is expected to impart profound changes to the configuration of manufacturing companies with regards to what their value proposition will be and how their production network, supplier base and customer interfaces will develop [16]. Amenities and services in the post-COVID era are hugely dependent on technology. Automation (industrial) refers to the processing of information and controlling the entire process remotely, without human interference. A target to meet every unit time is set for the machines and they are programmed accordingly to get results. Extensive use of IT, and other related technologies like cloud computing, is seen in these sectors to detect, optimize, control, and regulate the entire manufacturing process. Several huge, high-power-consuming machines with a huge range of functionaries are used in these sectors and the demand for them is steeply increasing [17]. These have a high production capacity and can be functioned by technicians remotely to achieve the desired outputs. The growth of automation in industries, use of cloud computing in it and in the provision of services and its maintenance is highly talked about. Several technologies are being introduced which revolve around them. It has made it very important to seamlessly integrate the various sectors of the process as a whole and cloud technology plays a significant role in this.

With the governments imposing lockdown and stay indoor protocols across the globe due to the pandemic, offices and educational institutions saw a shift to work from home and study from home techniques from the regular in person attendance. This saw the legislative, executive, and judicial branches of governments adopt specific work cultures. While hearings and meetings were conducted over video calls, all paperwork shifted to digital platforms. This saw a massive demand for cloud services for storage and documentation. Handwritten paperwork across the governmental departments, administrative offices and education institutions have become negligible in the post-COVID era. This also saw a huge rise of the amount of data generated each day, number of services accessed and used simultaneously, number of technologies shifted to cloud platforms for the ease of access, and higher security for the aforementioned. Educational institutions saw teachers taking classes over the web. Assignments, tests, and other academic work were created, assigned, completed, and submitted using various online services. Several cloud service providers also extended their services at a reduced rate for students and introduced other "work from home" offers. Over the years, the development of urban areas has seen a growth in urban population. This has given rise to the smart city concept to enhance the potential of city-level governance [18, 19]. The smart city concept consists of technologies like IoT using sensors, processors, actuators, vehicles, mobiles, etc., to collect data and AI technologies to analyze, understand, and draw conclusions from that data to further predict trends for the improvisation of planning, governance, and administration [20-23]. All these data collected encompass various big data methods and are stored, shared, and accessed using several cloud services.

There was a huge shift in the entertainment and media industry due to the effect of COVID-19. With people staying indoors mandatorily, cinema halls and theaters did not open to the public for quite some time, and a lot of productions were released in OTT platforms and other streaming websites. Theater and musical events were streamed live whereas movies and videos were launched online. This contributed to a large amount of this content being stored in the cloud for easy, fast, and safe access mostly on a subscription basis. Music, videos, news, theater performances, musicals, games, seminars, discussions, and interviews mostly contributed to the rise of using cloud-based services in the entertainment and media sector. Streaming apps and websites gained massive popularity and so did the cloud services which encompassed them. The food industry almost came to a standstill during the onset of this pandemic. But this issue was also resolved by using safe delivery methods. People preferred delivery and takeaway to dining out and this stopped the food industry from spiralling further down the financial pole.

Cloud services also collaborated with other ICT arenas, including but not limited to IoT, AI, ML, data analytics, robotics, business intelligence, and HCI, post pandemic. The Cloud of Things (cloud computing with IoT), also referred to as CloudIoT, saw some major new developments post pandemic which are expected to disrupt both the current and future internet [24]. While CloudIoT is predicted to create more business opportunities, it also possesses more threats and privacy concerns [25]. Providing quality service using CoT is another issue that has been widely talked about [26]. It is not entirely plausible to conclude the effects and usage of cloud-based services, its growth and a future path in and after the NCov era, but it is fair to state that cloud-based services have permeated across sectors and industries and brought about a revolution in how and where technology is used, thereby changing every aspect of our day-to-day lives.

4.2 Cloud-Based Solutions

4.2.1 Information and Communications Technology

IoT and CoT: The internet of things revolutionized the way the internet was used in our daily lives. It became an immersive experience and shifted our perspective of how we use it. From just viewing the internet as a means to access the world wide web and use its limited services, IoT ushered in the phase of technological remolding. With cloud technology taking off, the internet of things combined with the former to what we now call the Cloud of Things or simply CloudIoT. This amalgamation ushered in a variety of new technologies, services, challenges, and issues [25]. Moreover, it brought in "things" as a service to the previously four separate service sectors [25]. While several cloud service providers have shown exceptional promise, only a few are famously known and used. Several other new technologies have been created or are being developed, some of whose literature has been reviewed and summarized in Table 4.1.

4.2.2 Artificial Intelligence and Machine Learning

The shift to cloud-based platforms saw an evolution of easily available AI-based services with a focus on ML tools. Though these sectors overlap several of the arenas, visualization tools specifically for image processing and development tools for coders saw a major evolution with respect to ICT. Cloud-based image processing services provide high-quality images which are used across several industries [37]. With development and research shifting to a computational process with a huge amount of data available for research purposes, cloud-based developer tools have become a robust way for people from across the globe to come together and collaborate easily over the internet [38]. Some of the existing pioneers in these technologies which gained popularity post pandemic are listed below in Table 4.2.

Table 4.1: Information and communications technologies.

Reference	Service/Technology used	Field/Characteristics
[25]	Ambient assisted living	Betterment of patients who are disabled or diagnosed with chronic illness(es).
[25]	Environmental monitoring	Monitoring several levels of the environment as a whole, including but not limited to water levels, pollutant/toxins/gas contents of the air, fertilizers/nutrient levels in soil, conditions of environmental degradation, shift in static structures, and fire detection.
[27]	Intelligent Transportation Services	Installing monitoring devices in various places multiple locations to predict traffic, adjust signals, GPS, vehicle-to-vehicle/infrastructure communication to ultimately achieve overall city traffic control.
[27]	Large-scale emergency response services	Reported to all emergency centres from executives, public, various types of sensors, satellites, regarding findings, movement/location to assist in formulating relief plans and assigning and aiding first responders and other emergency departments and personnel.
[28]	Intelligent parking service	Consists of sensors, communicators, and apps to give information regarding the optimum available parking slot(s).
[29]	Control services for smart grids	Methods for conserving and managing energy and cost optimization using cloud services and store for further evaluation.
[27, 30-32]	Energy services related to smart buildings	Control HVAC, lighting systems and home and office appliances, using the wide range of given settings, evaluation methods for energy conservation and energy diagnostics services for lone or multiple similar buildings.
[27, 33]	Smart water networks	Early warning systems for faults in the entire network, real-time monitoring capabilities for water quality, force, and quantity, and presence of pollutants and/or toxins alongwith the location of the faults.
[27, 34]	Crowd control services	Track/study about the place, focus, and activity of the crowd and store the information for further prediction and analysis.
[25, 35]	Video surveillance	Video storage and processing collected using IP cameras and integrating VSaaS.
[36]	Smart logistics	Easy and automated management of goods shipped in between origin and delivery location

Table 4.2: Artificial intelligence and machine learning technologies.

Reference	Service/Technology used	Field/Characteristics
[39, 40, 41]	TeraRecon Inc	Provides trackable, access-controlled, and anonymous 3-D medical images to multiple users which are stored and processed using Cloud Technology.
[42]	Remote Sensing	Cloud-technology based operations conducted upon remotely sensed locational information provided using images and processed using geo-based image processing algorithms with an option of scalability.
[43]	Pelco Inc	Triple tiered cloud-based intelligent video surveillance
[44]	Target Recognition	Simple yet effective cloud-based SDK for mobile devices for target recognition in images
[45]	International Business Machines Corp	Uses responses regarding the image(s) or specific part(s) of the image(s) shared and stored in the cloud to train and test them in order to enhance image recognition using the classification model(s) and maintaining the score(s) over the cloud.
[46]	Red Hat Inc	Specializes in cloud-based software testing, evaluating, management, and development
[47]	Simulator	A cloud-based simulation environment for neural models.
[48]	Coderun	Cloud-based environment consisting of multiple pre-installed compilers for a wide variety of programming languages with access to source code using any device with internet connectivity.
[48]	Compilr (COMPILR)	Along with code completion and code sharing consists of various courses focusing on implementation.
[48]	jsFiddle (JSFIDDLE)	Cloud service for developers which includes multiple libraries and option to add external resources along with code sharing, options for HTML and CSS rendering, beautifiers, and pre-compilers, usually acts as the third-party run-time environment for tech giants of coding forums to test their JavaScript code.
[48,49]	Koding (KODING)	Cloud based development service which includes VM and SSH access along with complete guidelines regarding the testing, running, and deployment of codes and applications with a focus on bridging the gap between the developers and the system admins.
[48]	Cloud9 (CLOUD9)	Enhanced cloud programming environment including complete development solutions for businesses apart from the tools and infrastructure for learners, browser compatibility testing and live preview.
[48]	Creately	Collaborative cloud-based business and technical data visualization service with a built-in mock editor for UX design with a focus on integration facilitated by the provision of plugins for other platforms.
[48]	Codenvy (CODENVY)	Based on the open-source Eclipse Che cloud IDE, is a cloud-based cloud programming environment specializing in DevOps with a focus on easy access and provision of VMs using Docker Containers.

Reference	Service/Technology used	Field/Characteristics
[48]	Codeanywhere (CA)	Cloud-based access controlled collaborative platform for developers with facilities like Live preview.
[48]	SourceForge (SF)	One of the first cloud-based source code repository and allows collaborative development.
[48]	Git (Lab/Hub)	Biggest version-controlled, open source, authenticated, LDAP, industry-standard publishing platform for web developers which also offers hosting services alongwith issue-tracker, collaborative wikis, and other code development spaces.
[48]	Bitbucket (BITBUCKET)	Cloud based platform which offers integrations with leading businesses like JIRA and Bamboo Continuous Integration and also accommodates inbuilt customisable notification and messaging facilities.
[48]	CloudForge (CLOUDFORGE)	Cloud-based repository which offers remote management and has collaborations with TeamForge and Basecamp amongst others.
[48]	Microsoft's CodePlex (CODEPLEX)	Environment to develop and share code which can further be used to download the entire production phase.
[48]	Code (GoogleCode)	Cloud based free shared workspace for development of open-source projects.
[48]	CodePlex	Cloud repository which provides a source-code-controlled base over other such services alongwith discussion platforms and issue tracking amongst others.
[48]	ProjectLocker's	Cloud repository with private enterprise grade source code hosting facilities and other incomparable safety and security features.
[48]	LaunchPad	Provides centralized service of cloud repository specializing on cross=project collab and uses the Bazaar version control system for hosting the user's code.
[48]	Visual Paradigm (VP)	Provides cloud-based complicated modelling tools for SDE and supports code generation, modification, and synchronization for a multitude of tenets.
[48]	EclipseIDE	Most commonly used development environment and the only composition tool which has managed online integration using the Orion Project by migrating to Cloud.
[48]	Cloud SW processing and documentation tools	Various companies provide these services of a one place online help and reference store.
[48]	Rackspace	Cloud-based SW management tool which allows users to monitor infrastructure and VMs in their data centers alongwith providing real-time updates.
[48]	VMware	Cloud-based SW management tool which allows users to monitor OS, software and middleware running in physical, virtual, and cloud environments.
[48]	Azure Preview Portal	Provides infrastructure and software monitoring of one's cloud.
[48]	Cloud Portofolio Management	Various companies provide scalable infrastructure and software monitoring tools.
[50]	Jupyter Notebook	Clou-based tool which functions as a virtual notebook.
[51]	Distill.Pub	Open accessed peer-reviewed platform for scientific journals.
[51]	Dendro Research Notebook	Platform based on Open Science which includes both data processing and visualization facilities.
[52, 53, 54]	R Shiny	Open-source interactive web platform for application development using R.
[55]	QUANTANT TECHNOLOGY Inc	Cloud-based image rendering facility for further medical imaging which promoting collaborative analysis and diagnosis without additional software, hardware, and/or plugins.
[37, 56]	TOMAAT	AI empowered 4-D image for medical analysis on cloud platform.
[37, 57]	Kraken	Uses public cloud infrastructure for lossless image compression/decompression alongwith image resizing and other facilities
[37, 58]	Cloudinary	Cloud based image manipulation service
[37, 59]	Akamai	Cloud-based image processing service which provides web-compatible, platform-specific end products.
[37, 60]	Pixboost	Unique cloud-based image processing service which provides services like real-time image scaling and URL/JS based integration amongst others.
[61]	Acquia	SaaS platform that provides products, services and technical support regarding Drupal.
[62]	Heroku	A cloud PaaS based service supporting multiple programming languages which is continually being developed and upgraded.
[63]	ServiceNow	Develops cloud-based stage to assist businesses in the management of digital workflows for enterprise operations.

4.2.3 Data Analytics and Business Intelligence

Cloud services have been a major aid in businesses and enterprises shifting their work online. These services are spread across a range of sectors in the domain of data analytics and business intelligence. There has been a demand for cloud-based storage services from both academicians and industrialists due to its effectiveness at a fairly low cost [69]. Cloud storage has gained enough prominence for the concept of Storage as a Service to emerge [69]. Apart from the traditional cloud services of data storage and ensuring security, these services encompass multiple tools, which on integration help enterprises to overcome a multitude of problems. From addressing customer grievances to automated production lines, messaging services and virtual agents, cloud services have expanded to dominate the entire business space. Customers who avail such services judge the quality of services based on various criteria [64]. Studies have concluded that these criteria change over the years [64]. Hence, it is not erroneous to state that several of the cloud service providers which gained prominence during the 2020 pandemic were due to the unforeseen circumstances and the tilt of customer demand to certain aspects of those amenities. Use of microservices has been proposed keeping in mind the requirements of security and privacy concerns of the enterprises and the users [65]. Blockchain-based data management services have also been researched, along with their benefits and limitations [66].

Table 4.3: Data analytics and business intelligence technologies.

Reference	Service/Technology used	Field/Characteristics
[68]	MEGA	Cloud based service which allows forensic recovery of files to help prevent cybercrimes.
[70]	Sunverge	Provides cloud-based green energy management services with a specific usage of SaaS and StaaS.
[71, 72]	Twilio	One of a kind programmable cloud communications company which allows users to implement it to make and receive phone calls.
[73]	Dropbox	Popular cloud-based file hosting service with its high-speed synchronization facility and data centre capability as a specialty.
[73]	Google Drive	Well-known file virtual cloud-based storage service with remarkable amount of storage.
[73]	iCloud	Used across all Apple devices is renowned for its extremely secure features.
[74]	CloudMe	Cloud-based file storage service which has a freemium business model and also provides SSL encryption.
[74]	Yunpan	Chinese cloud-based storage service provider with unlimited free storage.
[75, 77]	SpiderOak	Online collaborative zero-knowledge blockchain-based file hosting and cloud storage service specializing in a technology to combat any cyberattack.
[75]	JustCloud	Provides users with simple and automatic backup solutions to their cloud along with device tracking and an all-round technical support for easy recovery of lost files.
[75]	pCloud	Preferred by several MNCs, is a cloud-based storage service with a focus on easy collaboration and enhanced encryption.
[76]	ownCloud	Cloud-based storage service which provides exceptional remote collaboration.
[77]	OneDrive	Cloud-based storage service integrated with Windows 10's inbuilt offline file manager.
[77]	Box	One of the oldest cloud content management company which is supported by diverse mainstream apps.
[77]	nextCloud	Provides a free software to install a cloud storage service on the client's personal server for extremely fast and safe access.
[78]	BotSociety	Personalized cloud-based virtual agent
[79]	Chatfuel	Cloud-based chatbot platform for Messenger.
[80, 81]	Sync	Provides cloud storage with end-to-end encryption.
[80]	FlipDrive	Cloud-based file storage service with customizable user permissions and offline backup facility on the provider's end.
[81,82]	SugarSync	Cloud-based multi folder backup service which also facilitates remote wiping of data for extra security.
[81]	Tresorit	One of a kind CSE which uses AES-256 in cipher feedback mode for data encryption.

It cannot be asserted that the cloud services listed in Table 4.3 gained prominence post pandemic, but it is necessary to note that the need for the migration to cloud became more apparent [67]. There was a high demand for CSPs and lesser capabilities to accommodate the demands [67]. All major and minor providers saw an upsurge of the demand for their cloud services. Listed in Table 4.3 are such businesses which provided their services as solutions in the sector of data analytics and business intelligence.

4.3 Impact of Industry 4.0 in the Cloud Computing Industry

Industry 4.0 is yet to have an agreed upon single definition [83] but it is mostly referred to as the Fourth Industrial Revolution, which will bring about widespread digitalization. A considerable transformation of the lifestyle and work culture is predicted by researchers apace with sustainable opportunities for growth and development. Economic sustainability for enterprises is predicted to take a forefoot in the change that enterprises will encounter [83]. New technologies will assist in improvising efficiencies and investment in those will be advantageous [84]. But most of these technologies which have been recently developed to aid Industry 4.0 are intended for bigger corporations and large-scale industries [85]. Small- and medium-scale industries do not greatly benefit from these, either due to the lack of funds for availing these technologies and services or due to these industries not being targeted by the former. This is where cloud technology erupted to hold a strong grasp of their needs with its property of scalability. Cloud service providers offered their amenities at various scales and plans. With the process of rapid shifting to the online platform the cloud industry became an accessible in-demand tool at the time of change and gained massive popularity.

Cloud technology has assisted various other ICT industries to bring forth innumerable solutions in many fields. Most of these technologies have supported the shift to Industry 4.0 standards and regulations. Apart from ICT-centric organizations, other domains have gained from the convergence of technologies with cloud. Healthcare has cloud-dependent AI-based tools for speedy detection of diseases, IoT-powered labs, online data storage and reservation systems amongst others. On the other hand, agriculture has become a great contributor to the economy with the introduction of smart agricultural methods, most of which rely on several cloud-based technologies and services. The entertainment industry saw a rise in OTT platforms and a demand for them. It is noteworthy to mention that almost all these OTT service providers engage CSPs for an enormous amount of data storage. Cloud has thereby become the lead technical service provider in terms of demand, usage, and economic acquirement.

Smart sustainable manufacturing has become the central theme of Industry 4.0. Intricate networks which can also be managed remotely are used extensively in this process to oversee the workings and make the entire process accessible and transparent. Automated autonomous machines, including robots, are preferred over human workforce due to their continuous placid working capacities all around the clock. Automated processes also decrease the production time, monetary involvement, and direct supervision. They can be easily programmed to achieve the desired result with a small slot for errors. Industry 4.0 is expected to yield a great number of benefits, including but not limited to efficiency of businesses and its associates, quality of products and services, and flexibility of operation and conduct [86]. AI and IoT technologies which allow real-time transfer of data through cloud and other such networking techniques form the centerpiece of Industry 4.0 from a technical point of view [86]. Industry 4.0 also includes a great deal of human-machine interaction. Recruiters have fallen back on scientific, research-based technicalities for the recruitment of the workforce and resort to subsequent analysis of their behavior and conduct based on several such theories [87].

To summarize, Industry 4.0's central hub consists of four specific targets: To achieve greater connectivity allowing the ease of data transfer based upon which enterprises develop solutions and conduct business, this further results in a transparent arena of information sharing leading to a growing trust between the producer and the consumer in turn increasing the scale of business, which further demands secure models of communication which is achieved using cloud technology and other cyber security measures and finally includes superior technical support due to the "globally available network" and promotes decentralized resolutions [88-92].

4.3.1 Agriculture and Forestry

We have long moved on from the days when agriculture conjured up images of the plough and the crowbar in our minds. Now, agriculture involves heavy machinery, complicated remotely controllable devices, automated sensors and other techniques which results in an exceptionally high yield and fetches tremendous profits. These days technological advancements are seen being considerably implemented in agriculture and forestry (see Table 4.4).

Table 4.4: Exceptional contributions of cloud computing in the agriculture and forestry arenas.

Reference	Service/Technology used	Field/Characteristics
[93]	Agri Info	Autonomic system (QoS-aware) for providing agricultural information as a service
[94]	Precision Agriculture	CoT based application with a 3-tier process wherein information is collected by sensors, processed in cloud-based platforms in the back end and the analysis is sent back to the front-end.
[95]	Precision Agriculture on Orchards	Is a decision support cloud-based system with automation facilities which can collect data and analyse it on the basis of scenario using the customizable decision module and also let(s) the client(s) control the on-field devices remotely.
[96]	Smart Agricultural Model	CoT based application to provide information to farmers on various topics ranging from the amount of fertilizers to the demand of their crops and their value in the market.
[97]	Cloud-based ERP	Cloud-based externally hosted Resource planning platform to aid the entire populace involved in the agricultural sector by combining both scientific and traditional knowledge with a focus on Indian agriculture.
[98]	Agverdict Inc	Receives geographical location-based information about agricultural activities and evaluates it further to provide a report of further processes and procedures.
[99]	Agriculture monitoring system	Uses self-implemented cloud technology integrated with sensors including but not limited to 4duino to collect and transfer relevant information.
[100]	Sensing as a Service	Uses CoT and other ICT technologies to proposes sensing as a service in the field of Agriculture.
[101]	MGNREGA	Use cloud technology to enhance governmental operations and aid in the development of rural areas under the MGNREGA scheme by the Government of India
[102]	Climate-smart agriculture	CoT based tool to make quick but beneficial decisions to make agricultural activities more profitable.
[103]	Digital Reach in Rural India	Tool to make rural India more inclusive in the Internet using cloud technology along with other ICT methods and aide in the technological development.
[104]	Ubidots	Based on public cloud, it as an IoT as a service company which provides real time updates about the clients' agricultural space.
[104]	Thing Speak	Based on public cloud, apart from providing live feeds about the agricultural patches, also lets the clients collect data, analyse it and then act accordingly.
[104]	ThingWorx	Built on a private cloud it is a purpose built IIoT system for quick development and deployment of smart connected devices.
[104]	Xively	IoTaaS tool which provides a system to deploy IoT application son the cloud along with providing cloud-based API with an SDK to ease the process of DevOps.
[104]	Connectera	Based on private cloud it creates digital sensors that farmers can use to monitor their livestock's movement amongst other unique features.
[104]	Phytech	Based on plant sensors it provides growers with a decision support service to increase their yields and to optimize irrigation over a private cloud on a pay-as-you-go basis.

Cloud technology has been the foremost ICT tool which has greatly influenced the technological influenced practices in the aforementioned areas. Cloud and IoT, which is also referred to as CoT or CloudIoT, has shown promise in several of the methods using both of these, including sensors and actuators which can be controlled remotely or can be automated in accordance with various external factors, including weather, soil quality, plant health, etc.

Livestock tracking and management services are also provided by CSPs. Timber tracking has also been talked about in the woodcraft industry. Fisheries and other related departments extensively use cloud technology for storing the collected data and for other monitoring tools. Forest fires, deforestation, and illegal habitation in protected areas can also be monitored with real-time updates with the use of various cloud services. Cloud has thereby helped evolve a rigid system with inflexible methods and laborious monitoring and management, to a well-equipped process, and developing these industries and assisting in their rise in the economic sector. Mentioned in Table 4.4 are some of the exceptional contributions of cloud computing in the agriculture and forestry arenas.

4.3.2 Entertainment, Media, and Hospitality

The recent pandemic saw a steep rise in the demand for OTT platforms. With nationwide lockdowns, people were tied to their homes, which led to a demand for the in-home services of the entertainment and media industry. Most of them use cloud for storage of their curated content. Many small news agencies emerged during the duration of the outbreak with limited capital and investment and the scalable cloud storage models were the best way to go. With travel slowly resuming, the hospitality industry took a great deal of safety measures for their guests, including cloud-based menus, passkeys, etc., which were widely introduced and used. Some of the emerging technologies in these fields are listed in Table 4.5.

Table 4.5: Emerging technologies in the entertainment, media, and hospitality arenas.

Reference	Service/Technology used	Field/Characteristics
[105]	Cloud Gaming	Service applications that help to store the video games in cloud and client can access the games as audio/video streams
[106]	Signiant Inc	Provides cloud-based web user interface to exchange content with other users, systems, and applications.
[107]	Ittiam Systems	Cloud-based solutions provider of intelligent video technologies.
[108]	5G Media	Cloud-based integrated programmable service platform for the development, design and operations of media applications in 5G networks
[109]	Decision Support models	New approach involving cloud-technology to tackle the decision-making problem faced by tourists while selecting hotels online
[110]	SST	Cloud-based self-service technology for secure access control in hotels and resorts.
[111]	Mixed Reality Game	Cloud based mixed reality gaming reduce the processing load on the mixed reality devices such as rendering and computing.
[112]	Face Video Retrieval	Cloud-based video content analysis with a focus on face video retrieval

4.3.3 Robotics and Automation, Manufacturing and Maintenance

Most of the finished goods we as consumers purchase and use are made in semi-automated industries. Several of those processes are done by robotics. These advancements in manufacturing and maintenance have reduced the involvement of manpower, in turn reducing risks, errors, and production time. Some of the emerging cloud-based technologies in product manufacturing and maintenance are listed in Table 4.6.

Table 4.6: Emerging cloud-based technologies in product manufacturing and maintenance.

Reference	Service/Technology used	Field/Characteristics
[113]	Maintenance	Cloud-based product service system for condition-based preventive maintenance aided by a shop-floor monitoring service and AR.
[114]	Monitoring	Cloud based service oriented system which enables real-time machine availability and execution status monitoring during metal-cutting operations, both locally or remotely.
[115]	Automation	Cloud-DPP methodology to support parts with a combination of milling and turning features, and process planning for multi-tasking machining centres with special functionalities
[116]	Robotics	Minimise the robot energy consumption during assembly as a Cloud service
[117]	Manufacturing, Robotics	Cloud-based manufacturing system to support ubiquitous manufacturing across multiple levels with the support of manufacturing cloud and function blocks.
[118]	Manufacturing	Interoperable manufacturing perspective based on cloud manufacturing, to support two types of cloud users, i.e. customer user and enterprise user
[119]	Maintenance	Cloud services are applied in WEEE recovery and recycling processes by tracking and management services.
[120]	Manufacturing	A smart manufacturing platform- Advanced Manufacturing Cloud of Things (AMCoT) designed and implemented to realize the proposed five-stage approach of yield enhancement and assurance.
[121]	Automation	Cloud-based ubiquitous robotic systems for smart manufacturing of customized products.
[122]	Robotics	Cloud-based robotics for faster and more powerful computational capabilities through massively parallel computation and higher data storage facilities.
[123]	Robotics	Berkeley Robotics and Automation as a Service (Brass), a RAaaS prototype for facilitating Automation-as-a-Service.
[124]	Manufacturing	Decentralized cyber-physical systems (CPS), which enables manufacturers to utilize a cloud-based agent approach to create an intelligent collaborative environment for product creation.

4.4 Significance and Impact of Cloud in the Pandemic Outbreak

As previously mentioned, the cloud computing industry has contributed to a great extent in various fields. It has meted out the demand for storage and security along with providing up-to-date technology at an affordable price. This industry has evolved into a flexible space wherein it can accommodate small businesses to big multinational corporations. Cloud has significantly contributed to the healthcare and data management sectors during and post pandemic. From solutions for patients to solutions for corporations during these trying times, cloud technology has done it all. It is also necessary to mention that cloud has amalgamated with multiple sectors, including ML, AI, robotics, IoT amongst others. CoT has gained exceptional prominence and has provided many solutions to problems deemed critical after the pandemic outbreak.

4.5 Conclusion and Future Directions

Cloud has not only assisted in shifting the manufacturing process to an online remotely controllable platform but has also contributed greatly to business automation. The fanfare regarding Industry 4.0 has grown since its introduction and this has seen a surge in the demand for a new generation of automation systems [125]. Several factories are making crucial adjustments in their industries in preparation for the wave of change that the concept of "factories of the future" has brought [126]. Cloud computing technology (CCT) is a revolutionary new way of leveraging the power of the internet to provide software and infrastructure solutions to businesses around the world [127]. Be it for live feeds or remotely accessible information regarding the entire production line, clients of cloud technology have got it all at their fingertips. Cloud came as a boon during the pandemic when global lockdowns were in place. This saw a trend of working remotely and collaborations across the globe. Once the benefits of integrating cloud in their businesses were realized, multiple businesses shifted to it. The four emerging technologies of AI, blockchain, cloud and data analytics together show great promise in digital business transformations [128]. Studies have shown that the aforementioned ABCD hybridization has a wide reach in diverse applications across a variety of vertical sectors in the post-pandemic era [128-130]. Thus, cloud technology is seen to be a front-runner in matters of business automation and it has significant potential and scope to further revolutionize the way we do business post pandemic.

References

1. Alashhab, Z. R., Anbar, M., Singh, M. M., Leau, Y. B., Al-Sai, Z. A., & Alhayja'a, S. A. (2021). Impact of coronavirus pandemic crisis on technologies and cloud computing applications. *Journal of Electronic Science and Technology*, 19(1), 100059.
2. Zeng, J., Huang, J., & Pan, L. (2020). How to balance acute myocardial infarction and COVID-19: the protocols from Sichuan Provincial People's Hospital. *Intensive Care Medicine*, 46(6), 1111-1113.
3. Manogaran, G., Thota, C., Lopez, D., & Sundarasekar, R. (2017). Big data security intelligence for healthcare industry 4.0. In *Cybersecurity for Industry* 4.0 (pp. 103-126). Springer, Cham.
4. Pavón-Pulido, N., López-Riquelme, J. A., Torres, R., Morais, R., & Pastor, J. A. (2017). New trends in precision agriculture: A novel cloud-based system for enabling data storage and agricultural task planning and automation. *Precision Agriculture*, 18(6), 1038-1068.
5. López-Riquelme, J. A., Pavón-Pulido, N., Navarro-Hellín, H., Soto-Valles, F., & Torres-Sánchez, R. (2017). A software architecture based on FIWARE cloud for Precision Agriculture. *Agricultural Water Management*, 183, 123-135.
6. Rajeswari, S., Suthendran, K., & Rajakumar, K. (2017, June). A smart agricultural model by integrating IoT, mobile and cloud-based big data analytics. In *2017 International Conference on Intelligent Computing and Control (I2C2)* (pp. 1-5). IEEE.
7. Vaishya, R., Haleem, A., Vaish, A., & Javaid, M. (2020). Emerging technologies to combat the COVID-19 pandemic. *Journal of Clinical and Experimental Hepatology*, 10(4), 409-411.
8. Kuo, M. H. (2011). Opportunities and challenges of cloud computing to improve health care services. *Journal of Medical Internet Research*, 13(3), e1867.
9. Javaid, M., Haleem, A., Vaishya, R., Bahl, S., Suman, R., & Vaish, A. (2020). Industry 4.0 technologies and their applications in fighting COVID-19 pandemic. *Diabetes & Metabolic Syndrome: Clinical Research & Reviews*, 14(4), 419-422.
10. Gachet, D., de Buenaga, M., Aparicio, F., & Padrón, V. (2012, July). Integrating internet of things and cloud computing for health services provisioning: The virtual cloud carer project. In *2012 Sixth International Conference on Innovative Mobile and Internet Services in Ubiquitous Computing* (pp. 918-921). IEEE. DOI: 10.1109/imis.2012.25.
11. Löhr, H., Sadeghi, A. R., & Winandy, M. (2010, November). Securing the e-health cloud. In *Proceedings of the 1st ACM International health Informatics Symposium* (pp. 220-229).
12. Ramgovind, S., Eloff, M. M., & Smith, E. (2010, August). The management of security in cloud computing. In *2010 Information Security for South Africa* (pp. 1-7). IEEE.
13. Brodkin, J. (2008). Gartner: Seven cloud-computing security risks Data integrity, recovery, privacy and regulatory compliance are key issues to consider. *Network World*.
14. Shacham, H., Waters, B. (2008, December). Compact proofs of retrievability. In *International Conference on the Theory and Application of Cryptology and Information Security* (pp. 90-107). Springer, Berlin, Heidelberg.

15. Wang, Q., Wang, C., Li, J., Ren, K., & Lou, W. (2009, September). Enabling public verifiability and data dynamics for storage security in cloud computing. In *European Symposium on Research in Computer Security* (pp. 355-370). Springer, Berlin, Heidelberg.
16. Culot, G., Orzes, G., Sartor, M., & Nassimbeni, G. (2020). The future of manufacturing: A Delphi-based scenario analysis on Industry 4.0. *Technological Forecasting and Social Change*, 157, 120092.
17. Li, P. (2019). Introductory Chapter: New Trends in Industrial Automation. In *New Trends in Industrial Automation*. IntechOpen. DOI: 10.5772/intechopen.84772.
18. Kummitha, R. K. R., & Crutzen, N. (2017). How do we understand smart cities? An evolutionary perspective. *Cities*, 67, 43-52.
19. Lee, J. H., Hancock, M. G., & Hu, M. C. (2014). Towards an effective framework for building smart cities: Lessons from Seoul and San Francisco. *Technological Forecasting and Social Change*, 89, 80-99.
20. Kankanhalli, A., Charalabidis, Y., & Mellouli, S. (2019). IoT and AI for smart government: A research agenda. *Government Information Quarterly*, 36(2), 304-309.
21. Mora, L., Deakin, M., Reid, A., & Angelidou, M. (2019). How to overcome the dichotomous nature of smart city research: Proposed methodology and results of a pilot study. *Journal of Urban Technology*, 26(2), 89-128.
22. Kummitha, R. K. R., & Crutzen, N. (2019). Smart cities and the citizen-driven internet of things: A qualitative inquiry into an emerging smart city. *Technological Forecasting and Social Change*, 140, 44-53.
23. Davenport, T. H., & Ronanki, R. (2018). Artificial intelligence for the real world. *Harvard Business Review*, 96(1), 108-116.
24. Botta, A., De Donato, W., Persico, V., & Pescapé, A. (2016). Integration of cloud computing and internet of things: a survey. *Future Generation Computer Systems*, 56, 684-700.
25. Aazam, M., Khan, I., Alsaffar, A. A., & Huh, E. N. (2014, January). Cloud of Things: Integrating Internet of Things and cloud computing and the issues involved. In *Proceedings of 2014 11th International Bhurban Conference on Applied Sciences & Technology (IBCAST)*, Islamabad, Pakistan, 14th-18th January, 2014 (pp. 414-419). IEEE.
26. Nawaz, F., Hussain, O., Hussain, F. K., Janjua, N. K., Saberi, M., & Chang, E. (2019). Proactive management of SLA violations by capturing relevant external events in a Cloud of Things environment. *Future Generation Computer Systems*, 95, 26-44.
27. Mohamed, N., Lazarova-Molnar, S., & Al-Jaroodi, J. (2017, April). Cloud of things: Optimizing smart city services. In *2017 7th International Conference on Modeling, Simulation, and Applied Optimization (ICMSAO)* (pp. 1-5). IEEE.
28. Ji, Z., Ganchev, I., O'Droma, M., Zhao, L., & Zhang, X. (2014). A cloud-based car parking middleware for IoT-based smart cities: Design and implementation. *Sensors*, 14(12), 22372-22393.
29. Bera, S., Misra, S., & Rodrigues, J. J. (2014). Cloud computing applications for smart grid: A survey. *IEEE Transactions on Parallel and Distributed Systems*, 26(5), 1477-1494.
30. Mohamed, N., Lazarova-Molnar, S., & Al-Jaroodi, J. (2016, March). CE-BEMS: A cloud-enabled building energy management system. In *2016 3rd MEC International Conference on Big Data and Smart City (ICBDSC)* (pp. 1-6). IEEE.
31. Mohamed, N., Lazarova-Molnar, S., & Al-Jaroodi, J. (2016, June). SBDaaS: Smart building diagnostics as a service on the cloud. In *2016 2nd International Conference on Intelligent Green Building and Smart Grid (IGBSG)* (pp. 1-6). IEEE.
32. Lazarova-Molnar, S., & Mohamed, N. (2017, March). Towards collaborative data analytics for smart buildings. In *International Conference on Information Science and Applications* (pp. 459-466). Springer, Singapore.

33. Kartakis, S., Abraham, E., & McCann, J. A. (2015, April). Waterbox: A testbed for monitoring and controlling smart water networks. In *Proceedings of the 1st ACM International Workshop on Cyber-Physical Systems for Smart Water Networks* (pp. 1-6).

34. Mohamed, N., & Al-Jaroodi, J. (2014, July). Real-time big data analytics: Applications and challenges. In *2014 International Conference on High Performance Computing & Simulation (HPCS)* (pp. 305-310). IEEE.

35. Prati, A., Vezzani, R., Fornaciari, M., & Cucchiara, R. (2013). Intelligent video surveillance as a service. In *Intelligent Multimedia Surveillance* (pp. 1-16). Springer, Berlin, Heidelberg.

36. Li, W., Zhong, Y., Wang, X., & Cao, Y. (2013). Resource virtualization and service selection in cloud logistics. *Journal of Network and Computer Applications*, 36(6), 1696-1704.

37. Pandey, N. K., & Diwakar, M. (2020, March). A Review on Cloud based Image Processing Services. In 2020 *7th International Conference on Computing for Sustainable Global Development (INDIACom)* (pp. 108-112). IEEE.

38. Randles, B. M., Pasquetto, I. V., Golshan, M. S., & Borgman, C. L. (2017, June). Using the Jupyter notebook as a tool for open science: An empirical study. In *2017 ACM/IEEE Joint Conference on Digital Libraries (JCDL)* (pp. 1-2). IEEE.

39. Zhao, T., Taylor, R. J., Li, G., Hu, S., & Wu, J. (2013). Cloud-based medical image processing system with anonymous data upload and download. *U.S. Patent No. US8553965B2*. Washington, DC: U.S. Patent and Trademark Office.

40. Zhao, T., Taylor, R. J., Li, G., Wu, J., & Jia, C. (2014). Cloud-based medical image processing system with access control. *U.S. Patent No. US8682049B2*. Washington, DC: U.S. Patent and Trademark Office.

41. Wu, Junnan, Robert James Taylor, Gang Li, Tiecheng Zhao, & Chunguang Jia. "Cloud-based medical image processing system with tracking capability." U.S. Patent No. US10078727B2, issued September 18, 2018.

42. Kang, S., & Lee, K. (2016). Auto-scaling of geo-based image processing in an OpenStack cloud computing environment. *Remote Sensing*, 8(8), 662.

43. Wang, L., Zhu, H., Aghdasi, F., & Millar, G. (2020). Cloud-based video surveillance management system. *U.S. Patent No. US10769913B2*. Washington, DC: U.S. Patent and Trademark Office.

44. Arshad, H., Lam, M. C., Obeidy, W. K., & Tan, S. Y. (2017). An efficient cloud based image target recognition SDK for mobile applications. *International Journal on Advanced Science, Engineering and Information Technology*, 7(2), 496-502.

45. Abedini, M., Von Cavallar, S., Chakravorty, R., Davis, M. A., & Garnavi, R. (2017). Cloud-based infrastructure for feedback-driven training and image recognition. *U.S. Patent No. US9760990B2*. Washington, DC: U.S. Patent and Trademark Office.

46. Ferris, J. M. (2009). Systems and methods for software test management in cloud-based network. *U.S. Patent Application No. US20090300423A1*. Washington, DC: U.S. Patent and Trademark Office.

47. Yamazaki, T., Ikeno, H., Okumura, Y., Satoh, S., Kamiyama, Y., Hirata, Y., ... & Usui, S. (2011). Simulation Platform: A cloud-based online simulation environment. *Neural Networks*, 24(7), 693-698.

48. Fylaktopoulos, G., Goumas, G., Skolarikis, M., Sotiropoulos, A., & Maglogiannis, I. (2016). An overview of platforms for cloud based development. *SpringerPlus*, 5(1), 1-13.

49. Mehta, N., & Gupta, V. K. (2013, November). A survey on use of SaaS of cloud in education. In *International Conference on Cloud, Big Data and Trust*, Bhopal, Madhya Pradesh, India (pp. 13-15).

50. Randles, B. M., Pasquetto, I. V., Golshan, M. S., & Borgman, C. L. (2017, June). Using the Jupyter notebook as a tool for open science: An empirical study. In *2017 ACM/IEEE Joint Conference on Digital Libraries (JCDL)* (pp. 1-2). IEEE.

51. Marques, B. M. (2020). Dendro Research Notebook: Interactive Scientific Visualizations for e-Science. Master in Informatics and Computing Engineering, FACULDADE DE ENGENHARIA DA UNIVERSIDADE DO PORTO.

52. Hanč, J., Štrauch, P., Paňková, E., & Hančová, M. (2020). Teachers' perception of Jupyter and R Shiny as digital tools for open education and science. *arXiv preprint arXiv:2007.11262*.

53. Beeley, C. (2016). *Web application development with R using Shiny*. Packt Publishing Ltd.

54. Sievert, C. (2020). *Interactive web-based data visualization with R, plotly, and shiny*. CRC Press.

55. Chen, Y., & Sunhwa, J. U. N. G. (2015). "Remote cloud based medical image sharing and rendering semi-automated or fully automated network and/or web-based, 3D and/or 4D imaging of anatomy for training, rehearsing and/or conducting medical procedures, using multiple standard X-ray and/or other imaging projections, without a need for special hardware and/or systems and/or pre-processing/analysis of a captured image data." U.S. Patent 10,734,116, issued August 4, 2020.

56. Milletari, F., Frei, J., Aboulatta, M., Vivar, G., & Ahmadi, S. A. (2018). Cloud deployment of high-resolution medical image analysis with TOMAAT. *IEEE Journal of Biomedical and Health Informatics*, 23(3), 969-977.

57. Chard, R., Madduri, R., Karonis, N. T., Chard, K., Duffin, K. L., Ordoñez, C. E., ... & Winans, J. (2015). Scalable pCT image reconstruction delivered as a cloud service. *IEEE Transactions on Cloud Computing*, 6(1), 182-195.

58. Fedorov, A., Beichel, R., Kalpathy-Cramer, J., Finet, J., Fillion-Robin, J. C., Pujol, S., ... & Kikinis, R. (2012). 3D Slicer as an image computing platform for the Quantitative Imaging Network. *Magnetic Resonance Imaging*, 30(9), 1323-1341.

59. Zhao, T., Taylor, R. J., Li, G., Wu, J., & Jia, C. (2014). *U.S. Patent No. 8,682,049*. Washington, DC: U.S. Patent and Trademark Office.

60. Mehrtash, A., Pesteie, M., Hetherington, J., Behringer, P. A., Kapur, T., Wells III, W. M., ... & Abolmaesumi, P. (2017, March) "DeepInfer: Open-source deep learning deployment toolkit for image-guided therapy." *Medical Imaging 2017: Image-Guided Procedures, Robotic Interventions, and Modeling* (Vol. 10135, pp. 410-416). SPIE. International Society for Optics and Photonics, 2017.

61. So, P. (2018). Software Development Kits and Reference Builds. In *Decoupled Drupal in Practice* (pp. 283-309). Apress, Berkeley, CA.

62. Middleton, Neil, and Richard Schneeman. Heroku: up and running: effortless application deployment and scaling. O'Reilly Media, Inc., 2013.

63. Alonso, R. A., Arneodo, G., Barring, O., Bonfillou, E., dos Santos, M. C., Dore, V., ... & Toteva, Z. (2014, June). Migration of the cern it data centre support system to servicenow. In *Journal of Physics: Conference Series* (Vol. 513, No. 6, p. 062032). IOP Publishing.

64. Lang, M., Wiesche, M., & Krcmar, H. (2018). Criteria for selecting cloud service providers: a Delphi study of quality-of-service attributes. *Information & Management*, 55(6), 746-758.

65. Esposito, C., Castiglione, A., Tudorica, C. A., & Pop, F. (2017). Security and privacy for cloud-based data management in the health network service chain: a microservice approach. *IEEE Communications Magazine*, 55(9), 102-108.

66. Zhu, L., Wu, Y., Gai, K., & Choo, K. K. R. (2019). Controllable and trustworthy blockchain-based cloud data management. *Future Generation Computer Systems*, 91, 527-535.

67. Athari, M. Migration Time Due to the Global Pandemic, an Assessment of Cloud Migration, Opportunities and Challenges (2020). In *Proceeding of the 2020 IEEE International Conference on Software Maintenance and Evolution (ICSME)*, Adelaide, SA, Australia.

68. Daryabar, F., Dehghantanha, A., & Choo, K. K. R. (2017). Cloud storage forensics: MEGA as a case study. *Australian Journal of Forensic Sciences*, 49(3), 344-357.

69. Ning, J., Huang, X., Susilo, W., Liang, K., Liu, X., & Zhang, Y. (2020, March-April). Dual access control for cloud-based data storage and sharing. *IEEE Transactions on Dependable and Secure Computing*, 19(2), 1036-1048.

70. Sanders, D., & Statman, S. (2018). Renewable energy integrated storage and generation systems, apparatus, and methods with cloud distributed energy management services. *U.S. Patent No. 9,960,637*. Washington, DC: U.S. Patent and Trademark Office.

71. Venkatesan, S., Jawahar, A., Varsha, S., & Roshne, N. (2017, November). Design and implementation of an automated security system using Twilio messaging service. In *2017 International Conference on Smart Cities, Automation & Intelligent Computing Systems (ICON-SONICS)* (pp. 59-63). IEEE.

72. Lawson, J., Wolthuis, J., Cooke, E., & Comer, J. (2014). System and method for enabling real-time eventing. *U.S. Patent No. 8,838,707*. Washington, DC: U.S. Patent and Trademark Office.

73. Yan, C. (2017). Cloud Storage Services. *Information Technology*.

74. Dehghantanha, A., & Dargahi, T. (2017). Residual Cloud Forensics: CloudMe and 360Yunpan as case studies. In *Contemporary Digital Forensic Investigations of Cloud and Mobile Applications* (pp. 247-283). Syngress.

75. Mohtasebi, S. H., Dehghantanha, A., & Choo, K. K. (2017). Cloud storage forensics: analysis of data remnants on SpiderOak, JustCloud, and pCloud. In *Contemporary Digital Forensic Investigations of Cloud and Mobile Applications* (pp. 205-246). Syngress.

76. Martini, B., & Choo, K. K. R. (2013). Cloud storage forensics: ownCloud as a case study. *Digital Investigation*, 10(4), 287-299.

77. KamalaKannan, T., Sharmila, K., Shanthi, M. C., & Devi, M. R. (2019). Study on Cloud Storage and its Issues in Cloud Computing. *International Journal of Management, Technology and Engineering*, 9(1), 976-981.

78. Ahmad, R., Siemon, D., & Robra-Bissantz, S. (2021, January). Communicating with Machines: Conversational Agents with Personality and the Role of Extraversion. In *Proceedings of the 54th Hawaii International Conference on System Sciences* (p. 4043).

79. Sarosa, M., Kusumawardani, M., Suyono, A., & Wijaya, M. H. (2020). Developing a social media-based Chatbot for English learning. In *IOP Conference Series: Materials Science and Engineering* (Vol. 732, No. 1, p. 012074). IOP Publishing.

80. Bhat, W. A., Jalal, M. F., Khan, S. S., Shah, F. F., & Wani, M. A. (2019). Forensic analysis of Sync. com and FlipDrive cloud applications on Android platform. *Forensic Science International*, 302, 109845.

81. Henziger, E., & Carlsson, N. (2019, December). The overhead of confidentiality and client-side encryption in cloud storage systems. In *Proceedings of the 12th IEEE/ACM International Conference on Utility and Cloud Computing* (pp. 209-217).

82. Shariati, M., Dehghantanha, A., & Choo, K. K. R. (2016). SugarSync forensic analysis. *Australian Journal of Forensic Sciences*, 48(1), 95-117.

83. Ghobakhloo, M. (2020). Industry 4.0, digitization, and opportunities for sustainability. *Journal of Cleaner Production*, 252, 119869.

84. Culot, G., Nassimbeni, G., Orzes, G., & Sartor, M. (2020). Behind the definition of Industry 4.0: Analysis and open questions. *International Journal of Production Economics*, 226, 107617.

85. Masood, T., & Sonntag, P. (2020). Industry 4.0: Adoption challenges and benefits for SMEs. *Computers in Industry*, 121, 103261.

86. Veile, J. W., Kiel, D., Müller, J. M., & Voigt, K. I. (2019). Lessons learned from Industry 4.0 implementation in the German manufacturing industry. *Journal of Manufacturing Technology Management*, 31(5), 977-997.

87. Oberer, B., & Erkollar, A. (2018). Leadership 4.0: Digital leaders in the age of industry 4.0. *International Journal of Organizational Leadership*.

88. Hermann, M., Pentek, T., & Otto, B. (2016, January). Design principles for industrie 4.0 scenarios. In *2016 49th Hawaii International Conference on System Sciences (HICSS)* (pp. 3928-3937). IEEE.

89. Bonner, M. (2017). What is industry 4.0 and what does it mean for my manufacturing?. Saint Claire Systems. Available online: https://blog. viscosity. com/blog/what-is-industry-4.0-and-what-does-it-mean-for-mymanufacturing (accessed on 29 December 2018).

90. Da Silva, F. M., Bártolo, H. M., Bártolo, P., Almendra, R., Roseta, F., Almeida, H. A., & Lemos, A. C. (Eds.). (2017). Challenges for Technology Innovation: An Agenda for the Future. In *Proceedings of the International Conference on Sustainable Smart Manufacturing (S2M 2016)*, October 20-22, 2016, Lisbon, Portugal. CRC Press.

91. Griffiths, F., & Ooi, M. (2018). The fourth industrial revolution-Industry 4.0 and IoT [Trends in Future I&M]. *IEEE Instrumentation & Measurement Magazine*, 21(6), 29-43.

92. Gronau, N., Grum, M., & Bender, B. (2016, July). Determining the optimal level of autonomy in cyber-physical production systems. In *2016 IEEE 14th International Conference on Industrial Informatics (INDIN)* (pp. 1293-1299). IEEE.

93. Singh, S., Chana, I., & Buyya, R. (2020). Agri-Info: cloud based autonomic system for delivering agriculture as a service. *Internet of Things*, 9, 100131.

94. Khattab, A., Abdelgawad, A., & Yelmarthi, K. (2016, December). Design and implementation of a cloud-based IoT scheme for precision agriculture. In *2016 28th International Conference on Microelectronics (ICM)* (pp. 201-204). IEEE.

95. Tan, L. (2016). Cloud-based decision support and automation for precision agriculture in orchards. *IFAC-PapersOnLine*, 49(16), 330-335.

96. Rajeswari, S., Suthendran, K., & Rajakumar, K. (2017, June). A smart agricultural model by integrating IoT, mobile and cloud-based big data analytics. In *2017 International Conference on Intelligent Computing and Control (I2C2)* (pp. 1-5). IEEE.

97. Ahmad, T., Ahmad, S., & Jamshed, M. (2015, October). A knowledge based Indian agriculture: With cloud ERP arrangement. In *2015 International Conference on Green Computing and Internet of Things (ICGCIoT)* (pp. 333-340). IEEE.

98. Wilbur, M., Ellsworth, J., Oommen, T., Mohapatra, A., & Thayer, D. (2017). Systems and methods for cloud-based agricultural data processing and management. *U.S. Patent No. 9,667,710*. Washington, DC: U.S. Patent and Trademark Office.

99. Adetunji, K. E., & Joseph, M. K. (2018, August). Development of a Cloud-based Monitoring System using 4duino: Applications in Agriculture. In *2018 International conference on advances in big data, computing and data communication systems (icABCD)* (pp. 4849-4854). IEEE.

100. Rao, B. P., Saluia, P., Sharma, N., Mittal, A., & Sharma, S. V. (2012, December). Cloud computing for Internet of Things & sensing based applications. In *2012 Sixth International Conference on Sensing Technology (ICST)* (pp. 374-380). IEEE.

101. Paul, P. K., and A. Bhuimali (2017). A novel approach and possibilities of cloud computing applications in the mgnrega: towards more social development powered by technologies. *Journal of Economic Research and Studies*, 2(2), 1-10.

102. Symeonaki, Eleni G., Konstantinos G. Arvanitis, and Dimitrios D. Piromalis (2017). Cloud computing for IoT applications in climate-smart agriculture: A review on the trends and challenges toward sustainability. In *International Conference on Information and Communication Technologies in Agriculture, Food & Environment*, pp. 147-167. Springer, Cham.

103. Zahoor, S., & Mir, R. N. (2019). IoT fog cloud model for digital reach in rural India. In *International Conference on Computer Networks and Communication Technologies* (pp. 717-725). Springer, Singapore.

104. Mekala, M. S., & Viswanathan, P. (2017, August). A Survey: Smart agriculture IoT with cloud computing. In *2017 International Conference on Microelectronic Devices, Circuits and Systems (ICMDCS)* (pp. 1-7). IEEE.

105. Nayak, P., & Sharma, S. K. (2017). Impact of Cloud Gaming in Health Care, Education, and Entertainment Services. In *Emerging Technologies and Applications for Cloud-Based Gaming* (pp. 261-283). IGI Global.

106. North, D., Vasile, T., & Clarkson, R. C. (2015). Systems and methods for secure cloud-based media file sharing. *U.S. Patent 9,830,330*, issued November 28, 2017.

107. Suresh, D., & Srinivasan, M. (2019). System and method for upload and synchronization of media content to cloud based media services. *U.S. Patent 10,380,077*, issued August 13, 2019.

108. Rizou, S., Athanasoulis, P., Andriani, P., Iadanza, F., Carrozzo, G., Breitgand, D., ... & Gordo, O. P. (2018, June). A service platform architecture enabling programmable edge-to-cloud virtualization for the 5G Media industry. In *2018 IEEE International Symposium on Broadband Multimedia Systems and Broadcasting (BMSB)* (pp. 1-6). IEEE.

109. Peng, H. G., Zhang, H. Y., & Wang, J. Q. (2018). Cloud decision support model for selecting hotels on TripAdvisor. com with probabilistic linguistic information. *International Journal of Hospitality Management*, 68, 124-138.

110. Anitha, M., Babel, A., Kumar, A., Rauniyar, A., & Zahid, K. (2019). Cloud-Based Secured QR Code for Self-service Access Control System at Resort and Hotels. In *Computing and Network Sustainability* (pp. 387-395). Springer, Singapore.

111. Dermawan, K. F., & Yusuf, R. (2020, December). Moving Mixed Reality Game to the Cloud: A Survey on Feasibility. In *2020 6th International Conference on Interactive Digital Media (ICIDM)* (pp. 1-4). IEEE.

112. Lin, F. C., Ngo, H. H., & Dow, C. R. (2020). A cloud-based face video retrieval system with deep learning. *The Journal of Supercomputing*, 76(11), 8473-8493.

113. Mourtzis, D., Vlachou, A., & Zogopoulos, V. (2017). Cloud-based augmented reality remote maintenance through shop-floor monitoring: a product-service system approach. *Journal of Manufacturing Science and Engineering*, 139(6), 061011.

114. Wang, L., & Wang, X. V. (2018). Machine Availability Monitoring and Process Planning. In *Cloud-Based Cyber-Physical Systems in Manufacturing* (pp. 83-103). Springer, Cham.

115. Wang, L., & Wang, X. V. (2018). Cloud-Enabled Distributed Process Planning. In *Cloud-Based Cyber-Physical Systems in Manufacturing* (pp. 105-123). Springer, Cham.

116. Wang, L., & Wang, X. V. (2018). Resource efficiency calculation as a cloud service. In *Cloud-based Cyber-physical Systems in Manufacturing* (pp. 195-209). Springer, Cham.

117. Wang, L., & Wang, X. V. (2018). Cloud robotics towards a CPS assembly system. In *Cloud-Based Cyber-Physical Systems in Manufacturing* (pp. 243-259). Springer, Cham.

118. Wang, L., & Wang, X. V. (2018). Architecture Design of Cloud CPS in Manufacturing. In *Cloud-Based Cyber-Physical Systems in Manufacturing* (pp. 297-323). Springer, Cham.

119. Wang, L., & Wang, X. V. (2018). Product Tracking and WEEE Management. In *Cloud-Based Cyber-Physical Systems in Manufacturing* (pp. 325-346). Springer, Cham.

120. Lin, Y. C., Hung, M. H., Huang, H. C., Chen, C. C., Yang, H. C., Hsieh, Y. S., & Cheng, F. T. (2017). Development of advanced manufacturing cloud of things (AMCoT)—A smart manufacturing platform. *IEEE Robotics and Automation Letters*, 2(3), 1809-1816.

121. Zhang, Z., Wang, X., Zhu, X., Cao, Q., & Tao, F. (2019). Cloud manufacturing paradigm with ubiquitous robotic system for product customization. *Robotics and Computer-integrated manufacturing*, 60, 12-22.

122. Saha, O., & Dasgupta, P. (2018). A comprehensive survey of recent trends in cloud robotics architectures and applications. *Robotics*, 7(3), 47.

123. Tian, N., Matl, M., Mahler, J., Zhou, Y. X., Staszak, S., Correa, C., ... & Goldberg, K. (2017, May). A cloud robot system using the dexterity network and berkeley robotics and automation as a service (brass). In *2017 IEEE International Conference on Robotics and Automation (ICRA)* (pp. 1615-1622). IEEE.

124. Zhang, Z., Li, X., Wang, X., & Cheng, H. (2017). Decentralized cyber-physical systems: A paradigm for cloud-based smart factory of Industry 4.0. In *Cybersecurity for Industry 4.0* (pp. 127-171). Springer, Cham.

125. Delsing, J. (2017). Local cloud internet of things automation: Technology and business model features of distributed internet of things automation solutions. *IEEE Industrial Electronics Magazine*, 11(4), 8-21.

126. Pei Breivold, H. (2020). Towards factories of the future: migration of industrial legacy automation systems in the cloud computing and Internet-of-things context. *Enterprise Information Systems*, 14(4), 542-562.

127. Attaran, M., & Woods, J. (2019). Cloud computing technology: improving small business performance using the Internet. *Journal of Small Business & Entrepreneurship*, 31(6), 495-519.

128. Puar, V. H., Bhatt, C. M., Hoang, D. M., & Le, D. N. (2018). Communication in internet of things. In *Information Systems Design and Intelligent Applications* (pp. 272-281). Springer, Singapore.

129. Singh, D. K., Sobti, R., Jain, A., Malik, P. K., & Le, D. N. (2022). LoRa based intelligent soil and weather condition monitoring with internet of things for precision agriculture in smart cities. *IET Communications*, 16(5), 604-618.

130. Akter, S., Michael, K., Uddin, M. R., McCarthy, G., & Rahman, M. (2020). Transforming business using digital innovations: The application of AI, blockchain, cloud and data analytics. *Annals of Operations Research*, 1-33.

5

Network Security in Evolving Networking Technologies: Developments and Future Directions

Uma Yadav[1], Ashish Sharma[2]

[1] RTM Nagpur University, Nagpur, India
[2]Computer Science & Engineering Dept, G H Raisoni College of Engineering, Nagpur, India
Email: uma.yadav12@gmail.com, ashishk.sharma@raisoni.net

Abstract

More than one computer system interconnected through wired or wireless links forms a network. Apart from a variety of networking protocols, networks broadly use client-server or peer-to-peer architecture for linked devices to communicate with each other. A networking subgroup includes network security. It necessitates safeguarding the network architecture that links the network's core to the network's edge. Network protection includes the implementation of IT security protocols and the deployment of network applications and hardware to protect the network, its infrastructure, and all of its traffic against unwanted cyberattacks, to defend against unauthorized access to all IT assets and resources available through the network, and to ensure that approved users have adequate access to all IT assets and resources in order to work effectively. The rapid expansion of the computer network infrastructure provides customers with excellent convenience as well as new security challenges. The topic of network protection typically consists of network infrastructure security and data security. It specifically relates to the network system's security, privacy, integrity and availability of data information in the system. The challenge of network protection persists across all levels of the data network, and the purpose of network security is to protect the secrecy, transparency, integrity, stability, usability and auditability of the network.

Keywords: Network security, cryptography, privacy, evolving security strategies

5.1 Introduction

Communication networks are used for a range of applications to transmit important and sensitive knowledge. As a result, individuals who plan to steal or exploit information or to interrupt or kill the structures that store or transmit it, are called to attention. The rapid growth of computer networks, especially the advent of the internet, is increasingly common and widespread for all kinds of information applications. However, all types of information that could be secretly exploited, captured, tampered with or destroyed by attackers for a variety of reasons are exchanged and processed in the public communication network, resulting in immeasurable damages. Network security services are primarily expressed in: unauthorized entry, claiming to be legitimate users, data loss, internet listening and network usage for virus transport, etc. With the increasingly prevalent issue of network protection, whether the network security problem can be solved has become one of the primary factors preventing network growth. The topic of network protection typically refers to network infrastructure security and data security. Network infrastructure protection is designed to deter illegal threats, access and degradation of the system, while data security is specifically intended to avoid misuse or illegal copying of private and sensitive data [1].

It is important to include highly technical functionality in a successful security policy. Protection, however, must start with more mundane concerns that are often overlooked: for instance, reducing physical access to houses, spaces, computer workstations, and taking into account the "messy" facets of human conduct, which can make any security precautions ineffective. In communication networks, the need for protection is not new. For example, in the late nineteenth century the American undertaker Almon Strowger realized that he was losing sales to his rival because the telephone operator was wrongly diverting calls from the recently bereaved to his competitor. Therefore, he developed a switching system that led to the first automatic telephone exchanges being implemented in 1897. This allowed users to use rotary dialing to make their own contacts to signal the appropriate destination.

Any of the key challenges involved in maintaining a fair degree of resistance to network threats will be studied here. Such attacks are organized and deliberately aimed, while others, arising from eavesdropping operations, may be opportunistic. Network security risks are increasingly evolving as vulnerabilities are found in both existing and newly installed programs, and solutions are needed to address those threats. Instead of comprehensive accounts of existing technologies, the research in this chapter can give you an insight into the more enduring network security concepts. The aim of this chapter is to identify some of the factors impacting network and data communications security and their consequences for users, and to implement some specific types of security services and their components, and to explain how they are used in networks.

Network security is a concern that has escalated due to the network's self-configuring and decentralized existence. Wireless sensor networks, mobile ad-hoc networks and vehicular ad-hoc networks are three types of ad-hoc networks. Because of the dissimilar kinds of active and passive attacks in the network, malicious nodes may enter the network. The passive kind of attack one in which the network output is not compromised by malicious nodes. The active type of attacks are those in which malicious nodes decrease the efficiency of networks in terms of different parameters. The different forms of active attacks that reduce the network are the black hole, wormhole, sinkhole, Sybil, etc. Different techniques have been proposed in recent times to detect malicious nodes from the network. In recent years, data protection methods and intrusion prevention mechanisms have been proposed to enhance the protection of the network. Stable channel establishment algorithms, which are Diffie-Helman, RSA, etc., are the field of research in network security that improves network security.

Network protection is the means by which such protocols and procedures prohibit any unwanted connection to a computer network. Network protection appears to include certain ways in which only approved users can access the network's data. Users are granted a special network access ID and password. In major organizations and entities, network protection is used to protect the network against any intrusion by third parties.

This chapter primarily details network security solutions, including authentication, data protection technology, intrusion detection system (IDS), and advanced network technology. Every network user has a network security problem, so we can place a high priority on network security, try to deter aggressive attacks and ensure network security.

5.2 Background on Attacks, Security Services and Challenges

5.2.1 Types of Attacks Possible on Network

A network assault is an attempt to obtain unauthorized access to an entity's network in order to steal data or engage in other destructive activity. As many individuals depend on the internet for different educational, personal and social tasks, the internet is today's world. The network relies on connectivity, data sharing, and company dealings or the whole exchange and business sector. Since the internet will connect networks to the whole world, certain people can try to hurt and disrupt these networks on a regular basis for various reasons. These intruders violate privacy by breaking into machines connected to the internet, either to gain access to data or to make it inoperable. In the face of a range of evolving ongoing network attacks and the threat of additional destructive future attacks, network security has featured prominently in the scope of computer networking. Passive and active attacks are two major attacks possible on the network.

- Passive Attack: Attackers gain access to a system and are able to detect or intercept sensitive data, but they leave the files unaffected.
- Active Attack: Not only can attackers obtain unauthorized entry, but they also alter content, either removing, encrypting or otherwise damaging it.

The following are popular threat vectors that can be used by attackers to infiltrate your network.

- Unauthorized Entry: Unauthorized access refers to attackers who gain access to a network without first receiving authorization. Unauthorized access attacks can be caused by weak passwords, a lack of security against social engineering, previously compromised accounts, and internal risks.
- DDoS Attack: Attackers build botnets, which are vast fleets of compromised machines that are used to deliver fake traffic to servers or network. A DDoS attack can occur at the network level, such as by sending massive numbers of SYN/ACC packets to overwhelm a server, or at the program level, such as by running complex SQL queries to put on a database.
- MIM Attack: The man-in-the-middle attack entails attackers intercepting traffic between your network and external sites or inside your network. If communication

protocols are not encrypted or attackers find a way to exploit encryption, they will intercept data being exchanged, collect user passwords, and hijack user sessions.

- Injection Attacks with Code and SQL: Often websites accept user inputs without validating or sanitizing those inputs. Attackers will then fill out a form or make an API call with malicious code instead of the intended data values. The code is run on the computer, allowing attackers to gain access to it.
- Escalation of Privilege: If an intruder has gained access to your network, they will use privilege escalation to extend their scope. Horizontal privilege escalation applies to attackers obtaining access to additional, nearby networks, while vertical privilege escalation refers to attackers gaining a higher degree of rights over the same systems.
- Insiders Threats: Malicious insiders with exclusive access to corporate networks will make a network highly insecure. Insider threats are difficult to detect and protect against because attackers do not need to access the network to cause damage.

Some common network attacks are:

1. Virus: A virus is not self-executing; it allows the user's presence to infect a computer and spread throughout the network. An example of this is an email with a malicious link or attachment. When a receiver opens the attachment or clicks the link, the malicious code is allowed, bypassing the device's security controls and rendering it inoperable. In this case, the customer inadvertently corrupts the device.
2. Malware: Without the support of the user, a worm can enter a device. An attacker on the same internet connection can send malware to that application when a user executes a vulnerable network application. The application may accept and execute the malware from the internet, thus creating a worm. This is much more successful than other forms of malicious content.
3. Worm: A worm can access a computer without the user's permission. When a user runs a compromised network program, an attacker on the same internet connection may deliver malware to that application. The application can accept and execute malware downloaded from the internet, resulting in the spread of a worm.
4. Phishing: The most common form of network attack is phishing. This entails sending emails from well-known tools or banks, as well as instilling a sense of urgency in customers to encourage them to act quickly. The email could contain a malicious link or attachment, or it could allow confidential information to be shared.
5. Botnet: Malicious ransomware has infiltrated this private data network. The hacker controls all of the computers on the network without the user's knowledge. Each machine on the network is known as a zombie because it is used to spread and infect a vast number of machines or as instructed by the attacker.
6. DoS Attack: A denial-of-service attack is a large-scale interference that destroys the victim's network or whole IT infrastructure, rendering it unavailable to lawful users. A DoS attack causes a device's system to become overwhelmed, preventing it from responding to service requests. A DoS attack is similar to a DDoS attack in that it targets the device's networks, but it's initiated from a large number of other host computers that have been infected by attacker-controlled malicious software. DoS attacks include the following:

a) Connection Flooding: By forming a large amount of TCP connections at the battered server, the intruder bogs down the host. The network is blocked by these bogus links and made inaccessible to legal users.

b) Vulnerability Attack: This involves sending a few well-crafted posts to the vulnerable device or software running on the intended server, to the point that the host fails, stops or worsens the service.

c) Bandwidth Flooding: The attacker prevents valid packets from reaching the server by transmitting a flood of packets. The packets sent are of such a large size that some are unable to bind to the target.

7. Distributed DoS Attack: This is a complex version of a DoS attack that is much more difficult to detect and defend than a DoS attack. In this attack, the attacker uses several compromised systems to target a single targeted system for DoS attacks. The DDoS assault leverages botnets as well.

8. Man-in-the-Middle Attack: This attack occurs when someone sits between you and the other person's conversation. By standing in the middle, the intruder captures, monitors, and effectively controls the touch. For example, as data is transmitted from the network's lower layer, the computers in that layer are unable to choose the recipient with which they exchange data.

9. Packet Sniffer: When a passive receiver is placed within the wireless transmitter's range, a copy of each transmitted packet is captured. These packets can include confidential information, essential and critical information, trade secrets, and other information that can pass through a packet receiver when flying over it. The packet's receiver would then serve as a sniffer, sniffing all of the transmitted packets that entered the package. The best defense against packet sniffers is cryptography.

10. DNS Spoofing: Hacking a computer involves corrupting data from the domain name system (DNS) and then loading it into the resolver's cache. This helps the name server to return an incorrect IP address.

11. IP Spoofing: This practice of using a false source address to inject packets into the internet is one of the ways to impersonate someone else. Anti-IP spoofing can be aided by end-point authentication that ensures the assurance of a response from the location we've decided.

12. Compromised Key: An attacker gains unauthorized access to secure communication by using a stolen key. A key is a hidden number or code that enables the sender or receiver to read confidential information without being notified. When an attacker gets access to a key and uses it to extract information, it is referred to as a compromised key.

5.2.2 Security Services

Any measure taken to protect the privacy and usability of data and networks is referred to as network security [3]. It encompasses both software and hardware. Connection to network infrastructure and facilities is regulated by effective network security. It identifies and restricts a wide range of attacks, preventing them from scattering or accessing the system. Security risks like personal data drip and economic espionage, identity fraud, and contamination of sensitive information networks have recently sparked widespread interest in the media and community. In general, computer network and information system protection shall have the following services [3]:

- Confidentiality: This confirms that knowledge is inarticulate in the circumstance that it is accessed by unauthorized person, systems, or organizations.
- Integrity: This ensures that the evidence has not been tampered with, either inadvertently or deliberately, by a third party.
- Authentication: This ensures that the data source is who it claims to be.
- Non-Repudiation: This confirms that the message's authorship cannot be disputed in the prospect.
- Availability: This ensures that the device facilities are accessible to legitimate customers.
- Privacy: This ensures that customer's characteristics are untraceable and unidentifiable based on their behavior and activities within the framework.

Several cryptographic tools have been established to alleviate various security extortions and ensure that the above-mentioned security services are provided. Table 5.1 shows some of the approaches being practiced.

Table 5.1: Security services and approaches.

Security services	Security Approaches	Examples
Confidentiality	Encryption and Decryption of message	Symmetric cryptographic algorithm (DES, CBC, AES, etc)and Asymmetric cryptographic algorithm (DSA, RSA, IBE, ABE, etc).
Integrity	Digest functions, message signature	Digest functions (MD5, SHA-1, SHA-256 etc), Message Verification Codes (MAC, HMAC)
Verification	Sequence of hash, Message Verification Code	HMAC, ECDSA, CBC-MAC,
Nonrepudiation	Data signature	HMAC , ECDSA
Availability	frequency hopping for Pseudo Random, Access control, firewalls, prevention systems for Intrusion	Signature Based system for Intrusion Detection, Statistical Anomaly Grounded system for intrusion detection
Secrecy	Unlinkability, Pseudonymity, k-anonymity, ZKP	DAA, EPID, Pedersen Commitment

5.2.3 Challenges

Recent IT skills are now becoming accustomed to increasing the consistency of the consumer service as well as the efficiency of essential applications in a number of fields. Communication networks allow the advancement of a wide range of technologies, including smart houses, hospitals, smart cities, smart grids and other manufacturing applications. Despite these challenges, communication networks' involvement as a critical component of the fundamental framework for distributing such delicate applications creates new privacy and security concerns.

The complexities of implementing network safety in emerging network structures and facilities are outlined and clarified in this section, as seen in Figure 5.1.

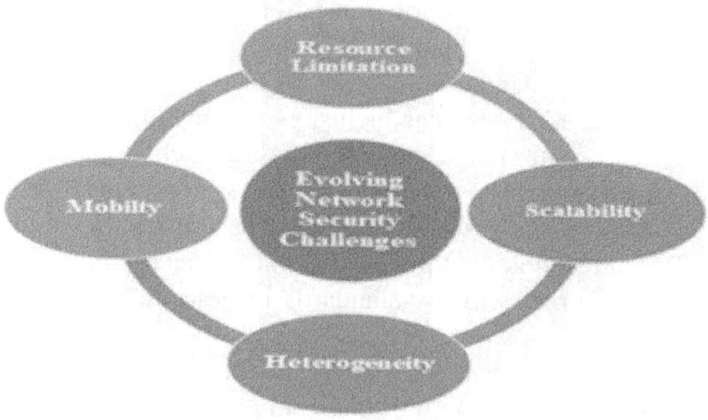

Figure 5.1: Security challenges.

- High Mobility: Implementing safe security systems, from integrated actuators and sensors from human bodies to smart cars, is a huge obstacle. Considering mobility in increasingly complex settings where network topology transitions often makes security solution implementation more difficult.
- Resource Restrictions: Most growing policies that participate in current networks, like embedded sensors and wearables, have limited memory, processing, and battery capacity. Since most cryptographic techniques are computationally costly, modifying them to guarantee a high degree of protection whilst reducing energy usage is a difficult and severe problem.
- Scalability: As the population's dependence on network-connected technology grows, the amount of smart devices continues to rise on a regular basis, creating yet another additional significant scalability problem for security solution growth.
- Heterogeneity: In a distributed networked environment, heterogeneity of connectivity protocols and information system architectures is a key problem in protecting the ecosystem. Connection between servers and sensor nodes or CPU items from numerous applications (which are varied in terms of calculation units and supply frequencies) is typically done over the internet, where communication mediums, networks, and protocols are all varied and devise different protection configurations. The variety of organizations participating in emerging networks creates a huge surface for assaults on all of them (e.g., DDoS attacks are unavoidable). As a result, designing (creating) an integrated protection approach that operates in heterogeneous environments is incredibly difficult.

5.3 Evolution of Network Security Strategies

The development of network security techniques is a product of the pervasive interconnectivity between computers, users, and scattered networks (i.e., an interactive environment, like the Internet of Things). In the networked ecosystem, conventional defense techniques as an alternative protecting a particular point within the system are becoming inadequate. Furthermore, some traditional security principles and best practices are inadequate in coping with the ecosystem's emerging security issues.

The standard security foundations of transparency, secrecy, and availability have all changed together with security techniques. However, it is becoming increasingly necessary to move beyond these standards in order to meet new requirements related to the physical climate, fitness, and safety. The inclusion of many integrated devices and utilities in the environment necessitates resolving crucial issues such as physical protection, disaster mitigation for smart or driverless vehicles, business continuity, wired HVAC systems, and online medical devices like infusion pumps and pacemakers, as well as interconnected city networks. Adopted technology policies dynamically incorporate certain security functionalities to meet the current security standards in order to mitigate emerging security threats.

Table 5.2 demonstrates further fields that are able to be integrated to improve the security in the current interacted ecosystem through simple security criteria.

Table 5.2: Additional evolving security requirements.

Functionality	**Description**
Proof of identity	Recognizing the risk level and current situation
Safety	Mitigating vulnerabilities and risks by implementing mitigation measures
Finding	Anomalies and events detection
Response	Response to events, prevention, and upgrades
Recovery	Continuous improvement over the life cycle

5.4 Different Evolving Security Approaches

In general, crucial problems, including resource restrictions and scalability, are still inconvenient in complex and contemporary networks, including vehicular networks, where the environment shifts frequently. Several portions of knowledge about the expedients' positions, battery-operated capacities, the amount of adjacent entities, and so on are often clustered together in the background. These pieces of data can be significant and therefore critical in improving safety, and they can also be used to design further modularity and security that is mindful of its surrounding strategies without relying on cryptographic methodologies. Consider the case of authenticating a single IoT computer A using a complicated cryptographic algorithm. It's often fascinating to avoid authenticating machine A with a cryptographic algorithm since it wants ample resources to complete the dense cryptographic methods and, as a consequence, saves its sequence when it's in a secure zone. It could be beneficial to consider other data correlated with system A in order to classify it deprived of relying on cryptographic policies. The details could include the period of its most recent authentication, A's venue, A's owner, and so on.

Figure 5.2 depicts the classification of evolving security approaches in the network.

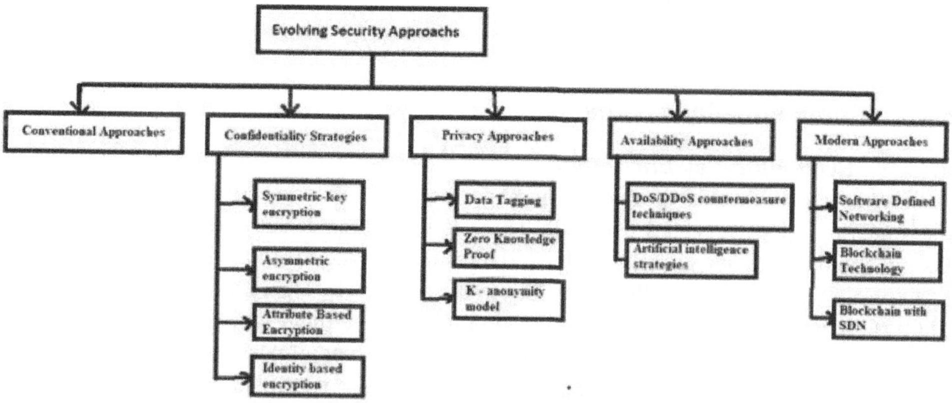

Figure 5.2: Different emergent security approaches.

5.4.1 Conventional Approaches

This classification includes cryptographic-based policies created specifically for the IoT, which is an evolving model that connects a variety of electronic policies and services. The main emphasis is on ensuring privacy, secrecy, and service accessibility. We need cryptographic mechanisms to protect data exchanged between objects in an emerging networked ecosystem from malicious hackers. As a result, only authorized users are permitted to view encrypted data. Cryptographic tools provide data confidentiality; however, in extreme cases, these trappings are inadequate or even incongruous in plans with limited resources. This is due to the nature of cryptographic procedures which necessarily want a proportion of computation and storage.

In a networked ecosystem, maintaining privacy is essential because data generated by smart objects is highly sensitive and inextricably linked to users' daily lives. The primary objective of privacy methods is to guarantee that the following standards are met:

- Inconspicuousness: This assures that a third party cannot distinguish the person's identity from that of other people in the organization.
- Unlinkability: This is the inability to conceal the individual's identity among the data they generate.
- Untraceability: This is the inability to trail activities and statistics emanating from a system individual's behavior.

The secrecy policies aim to protect delicate data while also introducing devices to conceal users' characteristics so that invaders are unable to track their actions.

Finally, one of the main intelligence services is the availability of network systems and services, which must be shielded against malicious assaults (like DoS or DDoS) or unintentional failures. The consequences of a breach of availability are commonly severe, ranging from financial losses (for example, in manufacturing systems) to safety issues (for example, in transportation schemes). Furthermore, guaranteeing availability has been a difficult job because attackers can disrupt the system by exploiting a wide variety of weaknesses at various levels (i.e., cryptographic algorithms, network design, software, etc.).

Most traditional security policies rely on central trusted organizations (i.e., in centralized environments) to ensure proper operation of security services.

5.4.2 Confidentiality Approaches

5.4.2.1 Symmetric-Key Encryption

In symmetric-key encryption the message is encrypted using a key, and the converted messages may be decrypted using the same key, making it easy to use but less secure. Using symmetric encryption algorithms, data is transferred in a way that cannot be read by someone who does not have the secret key to decode it. The algorithm reverses its behavior after the message has been delivered to the intended recipient who has the key, restoring the message to its initial and understandable state. A special password/code or a random string of letters or numbers produced by a protected random number generator may be used by both the sender and the receiver as the secret key (RNG). The symmetric keys needs to be generated using a RNG that is accredited according to industry standards, such as FIPS 140-2, for banking-grade encryption.

It therefore necessitates a secure method of passing the key from one party to the next. By forcing each system entity to exchange cryptographic secrets with a whole other system's entities, symmetric key strategies include anonymity. Symmetric centered cryptographic policies are beneficial in terms of effectiveness (due to the fact that they require less computation) and ease of implementation in hardware platforms. In reality, AES, DES, 3DES, IDEA, Blowfish, RC4, and RC5 are only a few examples. Despite the fact that symmetric key strategies save time and money, there are concerns regarding key management and scalability. The main distribution strategies used are either probabilistic or deterministic. To develop a full secure connectivity coverage in deterministic policies, each entity must form a protected link with all entities. While sharing a locked key of respective node in the network with all nodes is not guaranteed in probabilistic key distribution, nodes distribute keys to their neighbors based on certain probabilities, establishing secure routes among all units in the network.

5.4.2.2 Asymmetric Encryption

For encryption and decryption, asymmetric encryption employs a mathematically linked pair of keys: a public key and a private key. If the public key is used for encoding, the secrete key associated with it is used for decoding; if the private key is used for encoding, the public key associated with it is used for decoding. The sender and receiver are the two contributors in the uneven encryption workflow; each user has their own set of public and secrete keys. The source must first obtain the destination's public key. The sender then encrypts the plaintext (normal, readable text) with the destination's public key, resulting in ciphertext. The ciphertext is then sent to the recipient, who uses their private key to decrypt it and convert it to plaintext. One sender cannot read the communications of another sender due to the one-way environment of the encryption job, even though both have the receiver's public key.

Traditional asymmetric strategies require the authority to issue certificates to different system users and rely on public keys. It incorporates cryptosystems such as RSA, DSA, NTRU, and ECC, among others. Scalability, mobility, and key management effectiveness are the main benefits of asymmetric strategies. However, in terms of energy consumption, these techniques are ineffective for constrained devices. Although it requires further memory space for storing the keys, NTRU uses a much less computational uneven approach centered on the shortest path problem within a lattice [4].

5.4.2.3 Attribute-Based Encryption (ABE)

This is a promising modern cryptographic primitive for solving the issue of safe and fine-grained data exchange as well as decentralized access control. Via a policy admittance mechanism that defines relationships among attribute sets accustomed to encrypt records, attribute-based encryption (ABE) familiarizes a communicative way to manage private records accessibility. A key generation server (KGS) generates a private key for each valid user within the ABE scheme based on their attributes. A public key is often accustomed to encrypt records depending on a predefined plan. A legal user may only decrypt records if it possesses the attributes required by the regulation regulation; either a key policy ABE or ciphertext policy ABE may bc used.

- Key-Policy ABE: A significant class of ABE is key-policy attribute-based encryption (KP-ABE), in which ciphertexts are labeled with sets of attributes and private keys are paired with access mechanisms that govern which ciphertexts a user can decrypt [5]. KP-ABE has a lot of uses in data sharing on untrustworthy cloud servers. In most current KP-ABE systems, though, the ciphertext size increases linearly with the number of attributes contained in ciphertext. In KP-ABE, the records owner generates an entry erection A and encrypts the records with a series of features I. The consumer must then provide the characteristics that satisfy the access structure A in order to decode the ciphertext. In this way, a user may extract the private key for decoding the ciphertext [6].
- Ciphertext-Policy ABE: Data owners can exchange encrypted data using cloud storage with authenticated users while maintaining access control policies confidentiality using ciphertext-policy attribute-based encryption (CP-ABE) with private access control policy. However, a method to prohibit consumers from gaining successive access to a data owner's limited number of data items that raise a conflict of interest or whose mixture is vulnerable has yet to be investigated [7]. The encryption in CP-ABE is based on the admittance erection A. User A is only valid if he or she has a package of enough features I to satisfy the admittance erection (policy A) that has been added to the ciphertext [8].

5.4.2.4 Identity-Based Encryption (IBE)

This is a form of public-key encryption where the public key of a user is some unique information about the identity of the user (like an email address) and a reliable third-party server estimates the private key from the public key [9]. There's no reason to exchange public keys until you're able to share secure data in this way. The sender would quickly produce a public key and encrypt the data with the receiver's unique identifier. With the support of a trustworthy third-party server – the private-key generator – the recipient will produce the corresponding secret key (PKG). To use this encoding scheme, initially the PKG releases a principal public key and holds the principal secret key that corresponds (known as principal key).

Anyone can generate a public key equivalent to an identity by merging the principal public key with a well-known unique value given the principal public key (like an email address). To get the equivalent secret key, the owner of the individuality that produced the public key interacts with the PKG, which generates the corresponding secret key using its master secret key. Conversely, it necessitates the involvement of a reputable third party – the PKG.

Identity-based security tools effectively cope with scalability and intricacy by using an unforgeable series related to the user's identity (such as the consumer's phone number, email address, and so on) as the public key for data encoding, eliminating the need for certificates. The IBE approach is costly and consumes resources, and is not well suited for an emerging networked environment with a large number of devices that are resource constrained.

5.4.3 Privacy Approaches

5.4.3.1 Data Tagging

By associating bits of information (websites or images, for example) with tags or keywords, data tagging helps users to manage information more effectively. A non-hierarchical keyword or phrase assigned to a piece of information is called a tag (such as digital image, internet bookmark, or computer file). This type of metadata aids in the description of a product and enables it to be found again by browsing or scanning. Depending on the scheme, tags are selected informally and individually by the item's producer or viewer. The metadata that diverse data shares determines how well it can link and merge (even though it's co-located in the same data lake or cloud repository). Data tagging is just one facet of this, but it's a critical one.

Data tagging protects the anonymity of record flows by adding external identifiers, referred to as annotations, to record flows, allowing reliable computing institutions to interact with private data flows, hiding the identity of those responsible for the data [10]. However, depending on the size of the data, tagging mechanisms can experience computing issues. To demonstrate the ability to apply the tagging contrivance for programmable microcontroller (PIC) beneath restriction, the authors presented lightweight code models devoted to resource-constrained systems in [11].

5.4.3.2 Zero-Knowledge Proof (ZKP)

Data can be tested using zero-knowledge proofs (ZKPs) without disclosing the data. As a result, they have the ability to transform the way data is stored, used, and traded. A "verifier" and a "prover" are allocated to each transaction. The prover tries to show anything to the verifier using ZKPs without asking the verifier something more regarding the thing. The prover shows that they can compute anything without disclosing the input or the computing procedure by supplying the final result. The verifier, on the other hand, just knows about the performance.

A true ZKP must meet three criteria:

1. Completeness: It must persuade the verifier that the prover believes what they claim to know.

2. Soundness: If the knowledge is incorrect, it cannot persuade the verifier that the prover's information is correct.

3. Zero Knowledge(ness): It can tell the verifier little more [12].

Zero-knowledge proof (ZKP) is a useful tool for ensuring the protection of users' identities. The ZKP allows one party (prover) to validate any property to another revelry (verifier) by confirming its knowledge ownership without revealing it [13]. This idea is extremely useful in designing security mechanisms while preserving the integrity of users' data and resources. An assessment of certain ZKP protocols on ECC for resource-constrained devices was proposed in [13], which centered on the discrete logarithm problem. When ECC (having key length of 1024) is compared to RSA (for the same standard of security configuration), the results show that ECC (having key length of 1024) takes far less time and memory to execute. Notably, for small message sizes, the energy required for contact is negligible. Nonetheless, message fragmentation induces overloading within ZKP protocols above a certain threshold.

The $k-$anonymity model is considered a records holder such as the hospital or the bank that has a private database of person explicit, field structured records. Let's say the owner of the records wants to exchange a copy of it with scholars. How does a data owner issue a copy of their secret data that provides empirical assurances that the data subjects cannot be re-identified while the data is kept private? The approach involves a structured security paradigm called $k-$anonymity, as well as a series of implementation policies. If the information for each person found in the release cannot be separated from at least $k-1$ individuals whose information also appears in the release, the release offers $k-$anonymity security [14]. The $k-$anonymity security paradigm is notable because it acts as the base for the real-life systems Datafly, $k-$Similar and $\mu-$Argus, which all have privacy guarantees.

It is another theoretically promising approach for ensuring data protection in quickly emerging network systems, including IoT requests. Consider the case of a table containing a series of homogeneous data (including personal details such as ages, phone numbers, addresses, and so on). If the table field is a record of data that belongs to a single person, by shielding the owner's confidential details, the "$k-$anonymity" prototypes aim to hide each record inside the table and make it blurry from at tiniest $k-1$ records in the similar table [14]. The $k-$anonymity paradigm is often used in cloud and big data systems to secure the privacy of records streams provided by various users. Several attempts have been made to integrate $k-$anonymity prototypes in IoT implementations [15-17].

5.4.4 Availability Approaches

The DoS/DDoS defense mechanisms, for example, IP traceback approaches, are powerful mechanisms for detecting IP flooding and DoS attacks in real time on IP-based networks like the internet [25]. The distributed denial of service (DDoS) attacks have emerged as one of the most serious threats to computer network availability. In the internet, it's a tough challenge to tackle. Enhancing the stability of global network infrastructure is a necessary consideration in preventing a DoS attack. We discovered that the majority of methodologies focus on spotting and cleaning attack traffic close to the object. One of the most serious threats to network service providers and legal clients is a DDoS attack. Attackers are often looking for new ways to get into networks. The majority of protocols are vulnerable to DoS attacks. As a result, we must formulate an efficient automated approach to deter DDoS attacks by updating hardware and patching program vulnerabilities.

These approaches are primarily focused on improving the security of IP-based lightweight protocols that are mainly developed as variants of standard TCP/IP conventions in the emerging interacted environment, such as the internet of things. 6LoWPAN, DTLS 9 and IPv6 Routing Protocol for Low-Power and Lossy Networks (RPL) are just a few of the protocols commonly used to help safe end-to-end information sharing among the IoT devices while maintaining integrity and confidentiality [18]. These conventions, however, are not premeditated to protect against typical IP-centered DoS/DDoS attacks. Several safety resolutions have been researched in order to boost the RPL-centered 6LoWPAN routing protocol and the DTLS-centered transport layer's robustness and security against DoS attacks. IP routers and IoT gateways play the most critical function in current solutions, reviewing and evaluating envelopes in order to track malicious activity and take effective action [19].

Conversely, many security improvements of 6LoWPAN-centered IoT and RPL architectures are suggested at the network TCP/IP layer, especially at the routing side. The European ebbits project [20] proposed an interface to protect IoT devices based on 6LoWPAN from tampering and jamming attacks as well as DoS attacks [21,22]. They assisted in the development of an intrusion detection manager (IDM) that defends restricted devices from DoS attacks. They also have an intrusion detection system (IDS) architecture for monitoring 6LoWPAN packets, which raises warnings in the event of any misbehavior. The IDS works in promiscuous mode.

Artificial intelligence (AI)-based techniques, such as artificial neural networks (ANN), play a prominent role in the development of an effective intrusion detection system (IDS). For example, the authors of [23] looked at the usage of ANN in IoT to detect DOS attacks. They tested two forms of ANNs to see which one was more suitable for use as an IDS in emerging networked systems such as multilayer perceptron using normal weights and multilayer perceptron using limited weights. The findings show that while both ANN strategies decrease false positive recognition during the training phase, they ingest a lot of space, reducing their suitability in an environment with limited devices. Other researchers [24] looked at the possibilities of using Accumulative sum DDoS attack recognition in the context of emerging networks like IoT. The key goal of the cumulative sum (CUSUM) procedure is to perceive variations in the statistical process as record streams are generated in real time. DDoS is observed by network traffic analysis and computational computing. The algorithm constantly manages data and, in the end, detects variations that are related to some misbehavior in network traffic.

The author proposed [26] a novel artificial neural network-based approach to detect malicious network traffic that could be used in deep packet inspection-based intrusion detection systems. The proposed classification methodology is durable, reliable, and precise, according to the author. The innovative approach to malicious network traffic detection presented in this study has the ability to greatly increase the usefulness of intrusion detection technologies used in both standard network traffic analysis and network traffic analysis for cyber-physical systems like smart grids.

5.4.5 Modern Approaches

This group includes encryption strategies that are entirely focused on newly developed techniques rather than existing cryptographic technologies. In comparison to traditional methods, they are more suited to dealing with scalability problems. The ideas in this group are, on the whole, decentralized. Examples of two promising technologies are given below.

5.4.5.1 Software-Defined Networking (SDN)

This is a new networking architecture that separates network access from the data plane [29]. In a modern data center, it will make the system directly programmable, more versatile, and active to sustenance virtualized storage and server. Through decoupling and conceptualizing the control and records planes, SDN allows modular integration and advancement of modern networking applications. It has significantly altered the definition of networked services and the way they are designed and operated, as well as lowering the barricades to entry for different service providers. It is thought to be a likely alternative because of its size and diversity. Network administration is logically central in a software-driven controller in a standard SDN environment, while network devices such as switches can be called forwarding devices that can handle traffic based on specified flow tables [28].

With the support of predefined standards, the controller will customize these forwarding machines.

OpenFlow (OF) is the first unified communications protocol among the resistor layer and the structure layer in an software-defined networking (SDN) setting. It allows the centralized controller to manage switches without having to show any system source code [27]. In other words, it helps network managers to view and change physical and virtual switches and routers directly.

Consequently, SDN implementation in combination with network feature visualization (NFV) will improve resource usage in constrained devices within an advanced networked ecosystem. As a result, SDN offers a range of options for tackling some of the emerging problems of scalability, security, reliability, and QoS in a more scalable and effective manner [30].

The authors of [31] contributed a modern multi-domain SDN-centered IoT design that serves all networks with or deprived of infrastructure. They also developed a scattered security architecture to coordinate security policy across various SDN domains. The protection model grid, which is designed to address security heterogeneity problems, resolves tension issues caused by the execution of security policies through several realms. As a result, each SDN controller drives security policy within its field and coordinates through further SDN controllers outside the domain.

Challenges: The SDN-based technology approach mostly tackles security challenges in unified architecture activities.

As a result, centralized SDN controllers become particularly vulnerable to attacks. The biggest problem here is that centralized SDN controllers must be defended against emerging popular attacks like DDoS. Scalability problems arise when dealing with the vast quantity of policies in the essential data plan system. Furthermore, the southbound line between the data strategy and the SDN controller is a challenge to network performance reliability. For example, sited integrity issues within the OpenFlow protocol were discussed in [32].

In extremely complex settings, such as vehicular networks, where multiple messages are swapped between network topology and devices change frequently, a centralized SDN approach is ineffective. SDN techniques can take a long time to enforce security policies and configurations in such environments.

Real SDN strategies are more suited to certain implementations and work more effectively with problems such as quality of operation and heterogeneity. Nonetheless, in the majority of instances, their clustered architecture limits scalability.

5.4.5.2 Blockchain Technology

Blockchain technology has gained a lot of interest as a result of the success of bitcoin and other cryptocurrencies. It is a clever mix of many techniques, including peer-to-peer networks, distributed consensus protocols, cryptographic systems, smart contracts, distributed databases and game theory. It essentially facilitates mutual transactions between organizations (peer-to-peer design without relying on a central reliable server). It offers a decentralized method of establishing trust in economic and social activities, and hence has the potential to transform the future of commercial dealings, as well as the way computing and cooperation are carried out. Furthermore, there is no need for individuals to trust one another in order for it to function. It is almost impossible to challenge completed transactions once they have been authenticated using this technology. There are emerging security technologies that researchers have brought to light for the intent of mitigating safety threats in growing networks and facilities by offering security functionalities like data privacy, access management, and so on. In blockchain-based applications, a consensus process is used to reach an understanding about how to add or verify a data block.

Following are some vibrant examples of blockchain implementation within an emerging networked ecosystem:

1. IoT Blockchain Coalition (Guardtime based and Intrinsic ID based): Intrinsic ID is a company that offers cryptographic keys to authenticate surrounded devices using a technology called a physical unclonable function (PUF), which is mostly used to secure confidential activities like expenses and information related to government. Guardtime's objective is to provide a safety solution that is based on the keyless signature infrastructure (KSI) technology which includes a flexible blockchain result [33].

2. Chronicled: It is a recent business that offers blockchain-based applications. Its main emphasis is on addressing security issues, especially the authentication and recognition of IoT products. They claim that due to its tamper-resistant feature, blockchain could address a range of current security issues, especially now, where existing security tools like barcodes and QR codes can be easily forged.

In the advancement of networking services, blockchain technologies will offer certain benefits in the security domains [33]. Following are several blockchain structures that can be used to improve the safety of emerging system services:

- Transaction Security: All transactions are signed by the node before being submitted to the blockchain network, and they must be authenticated and checked by sappers. The dealings on the blockchain are nearly difficult to falsify or modify after they have been validated. This includes any evidence of traceable activities inside the system.

- Decentralization: Blockchain is favored as an effective encryption strategy within the environment due to the open architecture of emerging network systems and utilities. By preventing single points of failure, scalability accomplished by blockchain decentralized infrastructure will boost protection and increase toughness against DoS attacks.

- Pseudononymity: Blockchain aliases provide unlinkability between info and the identities of participating nodes. In blockchain, nodes are classified using shared keys.

Challenges: Despite the above-mentioned blockchain advantages, there are still a number of obstacles to overcome in order to implement blockchain technologies in current networks. Here are a few of the difficulties:

1. Computing and storage problems: Miniaturization of computers within an emerging networked environment limits capacities when it comes to storage and computation. As a result, the blockchain must be tailored to resolve the computing and storage concerns in order to meet the security needs of emerging networking technologies and services. The issue of adaptability is resolved in [33] by a proof-of-work (PoW) implementation, in which a new application layer is introduced solely to mask the blockchain information. In this way, machines with limited resources in a networked environment will engage without having to compute the PoW.

2. Time latency: If the same transaction confirmation cycle of 10 mins of the bitcoin-blockchain is implemented within emerging networking schemes and services, real-time applications may face security issues.

3. Scalability problem: Cisco estimated that there would be more than 20 billion linked IoT artefacts on the Internet by 2020 [34]. Despite the phenomenal popularity of the bitcoin blockchain, which has seen an unprecedented increase of customers over time, blockchain technology cannot assure scalability within a networked environment, including the internet of things.

4. Bandwidth usage: It is important to validate each of the devices' transactions generated bandwidth consumption problems due to the large amount of transactions caused by various devices in the interacted ecosystem.

5. Anonymity: Though it is difficult to deduce a person's identity from their pseudonym using blockchain transactions, this does not guarantee a truly anonymous transaction. This is because the peers in blockchain use pseudonyms that can be traced [35].

5.4.5.3 Blockchain with SDN

Even though SDN environments are similar, blockchain uses distributed controllers to increase the robustness of the system against single socket failure. The device layer and control plane can be condensed as two enormous components (referred to as the controller and application component) based on the specifications, where specific security protocols can be applied to provide protection. In the data plane, blockchain technologies can be customized to help improve the protection of multiple forwarding devices and distributed controllers. It is critical to design the configuration of programs, controllers, protection mechanisms, and related interactions in order to apply the interface.

DistBlockNet is a distributed IoT architecture introduced by Sharma *et al.* [36], which incorporates SDN and blockchain technologies. To counter threats, they created many protection measures such as the OrchApp and Shelter modules. The standard architecture of blockchain-centered SDN, referred to as DistBlockNet [36-40], will inherit many aids from both blockchains and SDN, including adaptability, availability, stability, scalability and protection.

While the combination of blockchain and SDN currently faces various security problems and concerns, they will complement each other in a number of realistic situations, either by adding SDN to blockchain or blockchain to SDN examples. More research on how to boost the safety and efficiency of blockchain-based SDN could be done in the future.

Challenges: Attackers may attempt to breach the network's security, credibility, and availability through leveraging SDN vulnerabilities. However, as a new concept, blockchain has a number of flaws that need to be addressed in reality, and it could become a tempting option for cybercriminals.

There remains a tremendous need to organize effective security measures to secure blockchain-based SDN against multiple threats due to various issues and challenges. It's also a smart idea to go with a private chain to make it more challenging for attackers. It is essential to deploy adequate protection measures and enforce safety policies such as flow and traffic management, policy compliance, and DoS defence to secure blockchain-based SDN in combative settings.

5.5 Discussion

In complex IoT contexts, meaning is crucial for properly solving security issues. Overall, the solutions in this group effectively satisfy efficiency criteria, including power consumption, computation, memory use, and service quality. These strategies, however, are less developed in the literature than other strategies, particularly in the sense of IoT. As a result, further technical activities should be directed toward filling the void and improving emerging technologies by using the context in which IoT devices emerge.

Despite the fact that many of the security issues confronting digital transformation are fresh, they can also be addressed by combining validated best practices with the introduction of a stronger security architecture. In order to protect highly distributed environments, high-speed authentication combined with monitoring is necessary. Internal segmentation is also intended to control and defend remote storage and interact by enforcing and organizing distributed and cloud-based security services that are capable of tracking and protecting information and devices throughout the network environment. Protection must connect the whole networked environment together. Security in emerging network structures and facilities necessitates automatic perceptibility from end-to-end points, as well as advanced monitoring technologies and hazard intelligence-driven transposition of reactions to mitigate threats at device pace.

What's needed is a framework-based, interconnected and dispersed security solution that can protect the entire networked environment, improve and assure durability, and safeguard figuring assets, which can be done via an interconnected, coordinated, and automated security infrastructure. This approach can efficiently track appropriate traffic, verify credentials and authentication, and enforce access control across the dispersed ecosystem.

5.6 Conclusion

A brief analysis of emerging network security techniques was presented in this chapter, as well as their challenges. In addition, key techniques for protecting the network environment were proposed. Due to the complexity of operating systems and service, the evolution of network security techniques will be a continuously growing and unpredictable paradigm. To address the emergent threats inside this complex environment, it is important to implement novel creative techniques and adopt the finest practices and standards in order to strengthen network security. The task of network protection occurs at all levels of the data network, and the aim of network security is to preserve the network's privacy, confidentiality, integrity, reliability, accessibility and auditability.

References

1. Bing, C., & Lisong, W. (2002). Research on architecture of network security [J]. *Computer Engineering and Applications*, 38(7), 138-140. DOI: 10.3321/j.issn:1002-8331.2002.07.047.
2. Ruambo, F. A. (2019). Network Security: A Brief Overview of Evolving Strategies and Challenges. *International Journal of Science and Research (IJSR)*, 8, 834-841. DOI: 10.21275/ART20194980.
3. Noura, H. (2016). Adaptation of cryptographic algorithms according to the applications requirements and limitations: Design, analyze and lessons learned. *HDR Dissertation, University of Pierre Marie Curie-Paris VI.*
4. Nguyen, K. T., Laurent, M., & Oualha, N. (2015). Survey on secure communication protocols for the Internet of Things. *Ad Hoc Networks*, 32, 17-31. DOI: 10.1016/j. adhoc.2015.01.006.
5. Wang, C. J., & Luo, J. F. (2012, November). A key-policy attribute-based encryption scheme with constant size ciphertext. In *2012 Eighth International Conference on Computational Intelligence and Security* (pp. 447-451). IEEE. DOI: 10.1109/CIS.2012.106.

6. Goyal, V., Pandey, O., Sahai, A., & Waters, B. (2006, October). Attribute-based encryption for fine-grained access control of encrypted data. In *Proceedings of the 13th ACM conference on Computer and communications security* (pp. 89-98).
7. Helil, N., & Rahman, K. (2017). CP-ABE access control scheme for sensitive data set constraint with hidden access policy and constraint policy. *Security and Communication Networks*, 2017. DOI: 10.1155/2017/2713595.
8. Bethencourt, J., Sahai, A., & Waters, B. (2007, May). Ciphertext-policy attribute-based encryption. In *2007 IEEE Symposium on Security and Privacy (SP'07)* (pp. 321-334). IEEE. DOI: 10.1109/SP.2007.11
9. Identity Based Encryption. https://doubleoctopus.com/security-wiki/encryption-and-cryptography/identity-based-encryption
10. Bruening, P. J., & Waterman, K. K. (2010). Data tagging for new information governance models. *IEEE Security & Privacy*, 8(5), 64-68. DOI: 10.1109/MSP.2010.147.
11. Evans, D., & Eyers, D. M. (2012, November). Efficient data tagging for managing privacy in the internet of things. *In 2012 IEEE International Conference on Green Computing and Communications (GreenCom)* (pp. 244-248). IEEE. DOI: 10.1109/GreenCom.2012.45
12. Maurer, U. (2009, June). Unifying zero-knowledge proofs of knowledge. In *International Conference on Cryptology in Africa* (pp. 272-286). Springer, Berlin, Heidelberg.
13. Chatzigiannakis, I., Pyrgelis, A., Spirakis, P. G., & Stamatiou, Y. C. (2011, October). Elliptic curve based zero knowledge proofs and their applicability on resource constrained devices. In *2011 IEEE eighth international conference on mobile ad-hoc and sensor systems* (pp. 715-720). IEEE.
14. Sweeney, L. (2002). k-anonymity: A model for protecting privacy. *International Journal of Uncertainty, Fuzziness and Knowledge-Based Systems*, 10(05), 557-570.
15. Huang, X., Fu, R., Chen, B., Zhang, T., & Roscoe, A. W. (2012, December). User interactive internet of things privacy preserved access control. In *2012 International Conference for Internet Technology And Secured Transactions* (pp. 597-602). IEEE.
16. Huo-wang, W., & Cheng, Z. H. O. N. G. (2013). Parallel clustering-based k-anonymity algorithm in internet of things. *Information Technology*, 12, 003.
17. Niu, B., Li, Q., Zhu, X., Cao, G., & Li, H. (2014, April). Achieving k-anonymity in privacy-aware location-based services. In *IEEE INFOCOM 2014-IEEE Conference on Computer Communications* (pp. 754-762). IEEE.
18. Sahraoui, S., & Bilami, A. (2015). Efficient HIP-based approach to ensure lightweight end-to-end security in the internet of things. *Computer Networks*, 91, 26-45. DOI: 10.1016/j.comnet.2015.08.002.
19. Kouicem, D. E., Bouabdallah, A., & Lakhlef, H. (2018). Internet of things security: A top-down survey. *Computer Networks*, 141, 199-221. DOI: 10.1016/j.comnet.2018.03.0 12
20. Kasinathan, P., Pastrone, C., Spirito, M. A., & Vinkovits, M. (2013, October). Denial-of-Service detection in 6LoWPAN based Internet of Things. In 2013 *IEEE 9th International Conference on Wireless and Mobile Computing, Networking and Communications (WiMob)* (pp. 600-607). IEEE.
21. Bhoyar, D. G., & Yadav, U. (2017, March). Review of jamming attack using game theory. In *2017 International Conference on Innovations in Information, Embedded and Communication Systems (ICIIECS)* (pp. 1-4). IEEE.
22. Bhoyar, D., & Yadav, U. (2017). A novel approach to detect jamming attack by means of game theory. *International Journal of Advanced Research in Computer Science and Software Engineering*, 7(4), 17-23. DOI: 10.23956/ijarcsse/V7I4/0110.

23. de Almeida, F. M., de RL Ribeiro, A., Moreno, E. D., & Montesco, C. A. (2016). Performance evaluation of an artificial neural network multilayer perceptron with limited weights for detecting denial of service attack on internet of things. *Traininge*, 11, 12.

24. Machaka, P., McDonald, A., Nelwamondo, F., & Bagula, A. (2015, November). Using the cumulative sum algorithm against distributed denial of service attacks in internet of things. In *International Conference on Context-Aware Systems and Applications (ICCASA)* (pp. 62-72). Springer, Cham.

25. Zeb, K., Baig, O., & Asif, M. K. (2015, March). DDoS attacks and countermeasures in cyberspace. In *2015 2nd World Symposium on Web Applications and Networking (WSWAN)* (pp. 1-6). IEEE. DOI: 10.1109/WSWAN.2015.7210322.

26. Shenfield, A., Day, D., & Ayesh, A. (2018). Intelligent intrusion detection systems using artificial neural networks. *ICT Express*, 4(2), 95-99. DOI: 10.1016/j.icte.2018.04.003.

27. Li, W., Meng, W., & Kwok, L. F. (2016). A survey on OpenFlow-based Software Defined Networks: Security challenges and countermeasures. *Journal of Network and Computer Applications*, 68, 126-139. DOI: 10.1016/j.jnca.2016.04.011.

28. Sahay, R., Meng, W., & Jensen, C. D. (2019). The application of software defined networking on securing computer networks: A survey. *Journal of Network and Computer Applications*, 131, 89-108. DOI: 10.1016/j.jnca.2019.01.019.

29. Bera, S., Misra, S., & Vasilakos, A. V. (2017). Software-defined networking for internet of things: A survey. *IEEE Internet of Things Journal*, 4(6), 1994-2008.

30. Hu, P. (2015, October). A system architecture for software-defined industrial Internet of Things. In *2015 IEEE International Conference on Ubiquitous Wireless Broadband (ICUWB)* (pp. 1-5). IEEE. DOI: 10.1109/ICUWB.2015.7324414.

31. Flauzac, O., González, C., Hachani, A., & Nolot, F. (2015, March). SDN based architecture for IoT and improvement of the security. In *2015 IEEE 29th International Conference on Advanced Information Networking and Applications Workshops* (pp. 688-693). IEEE. DOI: 10.1109/WAINA.2015.110.

32. McBride, M., Cohn, M., Deshpande, S., Kaushik, M., Mathews, M., & Nathan, S. (2013). SDN security considerations in the data center. *Open Networking Foundation-ONF SOLUTION BRIEF*, 15-16.

33. Conoscenti, M., Vetro, A., & De Martin, J. C. (2016, November). Blockchain for the Internet of Things: A systematic literature review. In *2016 IEEE/ACS 13th International Conference of Computer Systems and Applications (AICCSA)* (pp. 1-6). IEEE.

34. Evans, D. (2012). The internet of things how the next evolution of the internet is changing everything (april 2011). *White Paper by Cisco Internet Business Solutions Group (IBSG)*.

35. Brygier, J., & Oezer, M. (2016, January). Safety and security for the internet of things. In *8th European Congress on Embedded Real Time Software and Systems* (ERTS 2016).

36. Puar, V. H., Bhatt, C. M., Hoang, D. M., & Le, D. N. (2018). Communication in internet of things. In *Information Systems Design and Intelligent Applications* (pp. 272-281). Springer, Singapore.

37. Singh, D. K., Sobti, R., Jain, A., Malik, P. K., & Le, D. N. (2022). LoRa based intelligent soil and weather condition monitoring with internet of things for precision agriculture in smart cities. *IET Communications*, 16(5), 604-618.

38. Seth, B., Dalal, S., Le, D. N., Jaglan, V., Dahiya, N., Agrawal, A., ... & Verma, K. D. (2021). Secure Cloud Data Storage System Using Hybrid Paillier–Blowfish Algorithm. *CMC-Computers, Materials & Continua*, 67(1), 779-798.

39. Pramanik, S., Ghonge, M. M., Mangrulkar, R., & Le, D. N. (Eds.). (2022). *Cyber Security and Digital Forensics: Challenges and Future Trends*. John Wiley & Sons.

40. Sharma, P. K., Singh, S., Jeong, Y. S., & Park, J. H. (2017). Distblocknet: A distributed blockchains-based secure sdn architecture for iot networks. *IEEE Communications Magazine*, 55(9), 78-85. DOI: 10.1109/MCOM.2017.1700041.

6

The State of CDNs Today and What AI-Assisted CDN Means for the Future

Darothi Sarkar [1], Rana Majumdar[2], Dac-Nhuong Le[3]

[1] Institute of Engineering & Management, Kolkata, India
[2] Sister Nivedita University, Kolkata, India
[3] Faculy of Information Technology, Haiphong University, Haiphong, Vietnam
Email: darothisarkar@rediffmail.com, rana.majumdarwb@gmail.com, nhuongld@dhhp.edu.vn

Abstract

Replicating the network content in several strategically selected locations has become very popular because of its quality of experience. The content delivery network (CDN) brings data closer to the clients by deploying servers at the network edges, which eventually reduces the bandwidth constraint, packet loss and delay in response while gaining the clients confidence. Currently, Facebook, LinkedIn, Instagram and other social media are choosing OTT platforms that use CDN to cope with the increasing content demand, which used to be a major challenge before CDN. The first part of the chapter identifies the challenging areas involved in CDN on which its entire performance depends, namely server placement, content placement, request routing and load balancing. The second part of the chapter focuses on how artificial intelligence (AI) and machine learning (ML) are changing CDN. CDN keeps on using cutting-edge technologies to optimize response time with a disruption-free content delivery. The third part of the chapter emphasizes the role of emerging AI-based CDN during the pandemic, where the entire world shifted towards a virtual existence. Finally, the fourth part points out the drawbacks of CDN, as distributed denial-of-service (DDoS) attacks are a serious concern for CDN. Security problems along with how to address them are also discussed in this chapter.

Keywords: Content delivery network, surrogate placement, artificial intelligence, clustering technique, load balancing, fuzzy CDN, DDoS attack

6.1 Introduction

The gigantic amount of data generated by internet traffic is a major concern for today's web-based world. Placing the entire data in a single server results in delay in response and packet loss due to the bandwidth limitation. Therefore, CDN, an overlay network [1], has evolved with the aim of placing the content near to the client's proximity through content caching at network edges [2]. It was introduced by Akamai as an MIT research outcome [3]. The primary motivation behind this novel idea was to deliver the content with minimum latency and almost 100% availability. The first generation CDN was introduced to serve only downloadable files and static content. Similar to any other emerging technology, CDN has also gone through various stages of evolution. The second generation CDN incorporates dynamic content along with static. The third generation includes mobile data and rich media in it.

A traditional CDN is a combination of replica placement, content caching, content delivery, request routing, load sharing and accounting infrastructure. A typical content delivery infrastructure consists of one or more servers which contain the original content. These servers are termed as origin servers. The CDN service providers deploy a set of replica servers (also called surrogates or edges) in geographically dispersed points of presence over the entire network identified strategically [4]. Upon receiving a client request, the request routing infrastructure redirects it to an appropriate server without directing it towards the origin. The selection of the server depends upon various parameters like distance between the server and client, data availability, network traffic, server load, etc. The request routing component is also responsible for interacting with the distribution infrastructure to keep an up-to-date view of the content stored at the edges. The distribution infrastructure is responsible for replicating the content from the origin to the caches and ensures consistency of content at the edges. Maintaining client access logs and records of the CDN servers' usage is performed by the accounting infrastructure. This piece of information is used for traffic analysis and usage-based billing. CDN architecture along with its components is shown in Figure 6.1.

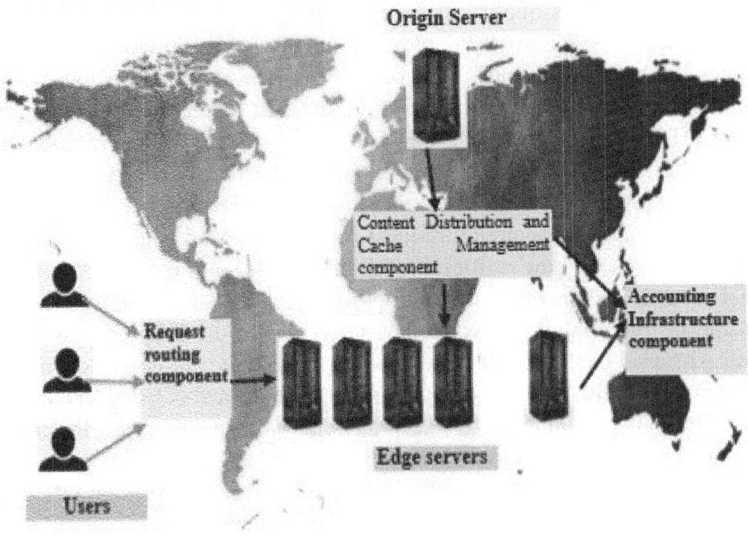

Figure 6.1: Content delivery network and its components.

The CDN has a vast application area. Most of the businesses having online presence tend to operate with CDN for all the services it provides like improved page load speed, reduced bandwidth consumption, handling high traffic, and load balancing between servers. The efficiency of a CDN can be evaluated based on some performance parameters like bandwidth, packet loss, latency, cache hit ratio, server utilization, CDN utility, etc. [3,5-6]. Sectors like advertising, mobile, media and entertainment, healthcare services, education, online gaming, e-commerce, government, etc., are realizing the benefits of CDN. YouTube, the most popular video generating site, was previously distributed by Limelight CDN, but after its acquisition, Google is using its internal CDN along with Akamai [6,7]. Leading CDN providers like Akamai and Limelight are distributing media for CNN and the BBC [6]. Social content delivery network (S-CDN) is an extended version of CDN through which huge content generated by social networking sites gets transmitted [7]. Recently, Telco CDN has been launched by telecommunication service providers that are managed by ISPs [8, 9].

This chapter has been divided into six sections. Section 6.1 introduces CDN and its architectural component along with the application areas. Section 6.2 describes CDN and its four major aspects and challenges. Section 6.3 considers CDN as an application area of artificial intelligence and machine learning. The role of CDN during a pandemic is discussed in Section 6.4. Section 6.5 identifies the security threats in CDN. Finally, Section 6.6 discusses the importance and need for CDN.

6.2 CDN and Its Challenges

The content delivery network (CDN) can be characterized by four major aspects, which are replica server placement, content replication, request routing and load balancing.

6.2.1 Replica Server Placement

A CDN provider needs to be prudent enough to place the surrogates over the network as efficiency of the network depends on the location as well as number of surrogates to a great extent. The main objective of CDN is to place the content closer to the clients. But placing too many servers increases the routing overhead as well as maintenance cost. So, optimizing the number of surrogates and identifying the best locations are an open research area in CDN. Surrogate placement can be considered as an NP-hard problem as it does not have any optimal solution [10]. A $k-$hierarchically well-separated tree ($k-$HST) and $k-$center problem are two graph-based approaches that can provide an optimal solution for replica placement [11]. But they aren't very popular because of their high complexity. Instead, a number of heuristic suboptimal replica placement algorithms have been introduced by several researchers. These algorithms are widely used for their relatively low computational complexity. These heuristic algorithms consider different optimization factors to improve their network efficiency. A greedy algorithm is one of the simplest approaches from an implementation point of view but its complexity is again a matter of concern [12]. In this approach, there are generally M iterations and in the first iteration the communication cost associated with all N potential sites gets analyzed and the node gets identified with the lowest cost. In the second iteration, the same process gets repeated for $N - 1$ sites and so on. A randomized algorithm is a similar type of approach where there are generally 10 iterations where in each iteration M sites randomly get selected out of N potential sites and associated cost gets analyzed. The iteration with minimum cost out of 10 repetitions is finalized [12]. HotSpot and HotZone are two popular approaches which take network load and traffic information under consideration to deploy surrogates in highly trafficked generated areas [12, 13]. There are some topology informed strategies available for deployment which generally sort all potential nodes in the network considering the network pattern [11]. One such approach is the Constrained Mirror Placement (CMP) that was introduced by the authors in [14], which considers the round-trip time (RTT). Max Fanout is an approach which identifies K sites with maximum outgoing edges for placing the surrogates [15]. In Flow Count analysis, the traffic passes through individual nodes and selects sites with maximum traffic flow as potential locations [16]. Another group of algorithms applies different clustering techniques to deploy servers. To find the best locations, GeoIP clustering technique applies subtractive and Fuzzy C-Means (FCM) clustering considering geographical details, including latitude and longitude extracted through IP addresses from the user log [17]. $k-$Means clustering is a popular approach used by several replica placement algorithms. NetClust is an algorithm that uses $k-$means and optimizes the network latency by an optimal matching between clients and servers [18]. A population-based clustering is proposed in [19], which also applies $k-$means to find the best locations for edge placement and also optimizes the number of servers based on cluster populations.

Table 6.1 shows different placement algorithms along with the parameters under consideration and their optimization factors [4, 20].

Table 6.1: Replica placement algorithms along with their optimization factors.

Algorithm	Parameters taken into consideration for designing the algorithm	Optimization Factors
Greedy	Network pattern as a graph and communication cost	Communication cost
Random	Network pattern as a graph and communication cost	Communication cost
HotSpot	Traffic information and network load	Latency, traffic load
HotZone	Traffic information and network load	Latency, traffic load
Max Fanout	Network pattern and number of outgoing edges	Average latency, traffic load
Flow Count	Network topology and traffic passes through individual node	Bandwidth, cross traffic
Constrained	Mirror placement network topology and RTT	RTT, cross traffic
GeoIP	Geographical details including latitude and longitude	Latency
NetClust	Network position and latency	Latency, Deployment cost
Population-based clustering	Cluster population	Server utilization factor

6.2.2 Content Replication

Once the servers are placed, the next important question is how to place the content in the surrogates. There are two standard ways by which content gets disseminated to the edges from the origin; one is push based and the other one is pull based. In Pull-CDN, upon receiving a user request, the surrogate retrieves the content from the origin and caches the content for the future [11]. Pull-CDN is either cooperative or non-cooperative. Cooperative Pull-CDN redirects a request to nearby surrogates in case of cache miss whereas non-cooperative directs the request to the origin server directly [21]. In Push-CDN, the origin server pushes content to all the surrogates and in absence of any requested content the server directs the request to the nearby replicas. As per the replication strategy, CDN follows either a full-site or partial-site approach [11, 22]. It can be interpreted from the name itself that full-site replication means replicating 100% content to all surrogates from the origin. On the other hand, a partial-site replicates selected content from the origin server. This content selection depends on several factors introduced in several approaches. Full-site replication is advantageous from an implementation point of view and also ensures almost 100% data availability. But the main drawback of this approach lies in the large storage requirements and frequent content update. Any update in origin should be reflected in all replicas. Partial replication on the contrary, replicates embedded objects as they get updated infrequently. Based on the approaches for object selection, partial-site replication gets divided into several categories. The content gets empirically selected by the website administrator in an empirical-based approach which is a heuristic one [11]. The popularity-based approach places the most popular content in the edges based on the statistics but it does not exhibit reliable performance as the popularity varies considerably and also the same is not available for newly added objects [23]. Object-based is a greedy approach where each content gets replicated in each server in the form of an object maintaining the storage constraint [24]. The iteration which gives the best performance gets selected. Instead of providing optimized performance, this approach is not adopted in practice because of its high complexity. Cluster-based is another approach where the web content gets replicated in clusters. Clusters are formed based on URL or user's sessions [25].

Table 6.2 shows the partial-site replication approaches along with their pros and cons.

Table 6.2: Partial-site replication approaches with their advantages and disadvantages.

Partial-site replication Strategies	Advantage	Disadvantage
Empirical-based	Easy implementation	Uncertainty in choosing the right heuristics
Popularity-based	Reduced latency for already existing content	Time consuming implementation, statistics are often not available for newly introduced content
Object-based	High performance	High complexity
Cluster-based	Reduce client download time and the load on server	Deployment complexity

6.2.3 Request Routing

Request routing can be considered as a crucial CDN component as it is responsible for directing the user request to the appropriate server, bypassing the traffic congestion in order to optimize the download time and response time. Routing of user requests gets directly impacted by content dissemination. Routing is easy in the case of full replication as there is less chance of cache miss. But request routing plays an important role in partial-site replication as hit ratio gets decreased in this approach. Request routing algorithms and routing mechanism are two components of CDN request routing [26]. Routing algorithms select an appropriate server against a client request and selected information goes to the client via routing mechanism. Based on the approaches followed for routing, all the routing algorithms have been divided into two broad categories: adaptive and non-adaptive request routing [27]. Non-adaptive routing doesn't consider the current network situation; rather it uses some heuristics for selection of surrogates. The simplest one is to use a round-robin approach where all the client requests get distributed to all the servers in the network in order to balance the load with a constraint of similar processing power for each surrogate. But this approach doesn't fit for a widely distributed network as client-server distance has not been included as a performance metric [28]. Another non-adaptive algorithm ranks all the servers based on the requests served by each so far and directs the client requests according to that but taking client-server distance as an influencing metric [29]. There is a non-adaptive algorithm that identifies the powerful servers based on the percentage of user requests received by the servers and directs client requests according to that to achieve better resource utilization [30]. Though non-adaptive algorithms are easy to implement, they are only efficient when the heuristics are met.

Adaptive algorithms select the surrogates based on present system state, including some factors like server load and network link congestion. Globule uses an adaptive approach which selects the closest replica server in terms of network proximity [31] where the path length gets updated periodically. Another such algorithm takes latency as a factor and from the user access log it identifies which server has the minimum latency and accordingly directs the user requests to that server [32]. Cisco DistributedDirector proposed an adaptive algorithm which considers a combination of three different metrics like intra-AS distance, inter-AS distance and end-to-end delay [30]. This algorithm has a drawback as it introduced additional traffic because of the latency measurement technique.

The features for both adaptive and non-adaptive algorithms are compared in Table 6.3.

Table 6.3: Comparison between adaptive and non-adaptive request routing algorithms.

Features	Non-adaptive request routing algorithm	Adaptive request routing algorithm
Approach	Uses heuristics	Considers current network status
Implementation Complexity	Easy implementation	Complexity gets increased with the ability to adapt to an enduring situation.
Robustness	Efficient only where the heuristics are satisfied.	Exhibits high system robustness

6.2.4 Load Balancing

Load balancing is another very crucial aspect in CDN in order to enhance the QoS and curtail the faults. Load balancing can be static or dynamic [33]. In static load balancing, prior to the execution the processes are assigned to the processors as the processors' performance is determined before the execution starts [34]. Round robin, where the client requests are evenly distributed to all the processors, is the simplest approach for static load sharing but it performs well with the constraint that the number of processors should be equal or more than the number of processes at any time [35]. Another such algorithm is central manager algorithm, where a central processor decides the host for a newly created process depending on the system load state information [36]. It may result in bottlenecks because of interprocess communications for load state sharing between all the servers and the central manager. According to the threshold algorithm, the server gets selected based on the processor state, which can be characterized as one of three levels: underloaded, medium and overloaded. This approach shows a significant disturbance in load sharing when all the local and remote processors are overloaded and a newly created process has to be assigned in an overloaded local server. In contrast, dynamic load balancing distributes the load during runtime. Central and local queue algorithms are standard ways of executing dynamic load distribution [37]. In the central queue algorithm, the cyclic queue in the main host buffers all the client requests and whenever it receives a request for new activity from any processor, it removes the first activity from the queue. Local queue algorithm introduced a threshold parameter that defines the minimum number of ready processes assigned to each processor [38]. Some recently proposed approaches use several parameters for highly dynamic distributed load balancing, which include fault rate, network load, response time, periodical status exchange information, global and local energy consumption, communication cost, etc. [39-41].

6.3 Importance of AI in CDN

All the aspects in CDN need some amount of intelligence to make a right decision. In this regard, the flavor of AI and ML has been blended along with CDN. Use of machine learning in CDN is an emerging open area to be explored. In replica server placment, upon receiving a client request, the request has to be redirected to an appropriate surrogate. To identify the right server, a number of algorithms have been proposed. In recent years, because of the gigantic amount of data generated by the web traffic, it is infeasible to handle them manually. This large amount of unstructured data can be transformed into a structure through predictive modeling [19]. $k-$Means is the most well-known clustering algorithm used widely for surrogate placement [42]. This clustering starts with random seeds with the goal of dividing all N potential sites into K clusters with a constraint $K << N$. All N sites are compared with each seed in terms of Euclidean distance and get assigned to the cluster with the nearest seed. The mean for each K cluster gets recalculated and again the same process gets repeated. The repetition stops once the mean values do not change or the changed values are almost zero.

The entire algorithm can be summarized in Algorithm 6.1.

Algorithm 6.1 $k-$Means clustering

Input:

N: Number of all potential sites in the network;

K: Number of clusters;

Output: K number of clusters with their centroid positions and cluster members.

BEGIN

Step 1: Initialize the seeds for K clusters.

Step 2: Assign each N node to a cluster with lowest Euclidean distance between the node and the centroid.

Step 3: The centroids of the clusters get modified.

Step 4: Steps 2 and 3 get repeated until the mean values do not change.

END

Apart from the $k-$means clustering, Fuzzy $C-$means (FCM) clustering is also used to find the appropriate locations for deploying servers [17]. FCM is applicable when the number of surrogates is known beforehand and this approach provides better results with overlapped data sets. On the contrary, subtractive clustering is used to find the deployment locations where the number of surrogates depends on the influence range given as an input to the algorithm [17].

Replication strategies are generally designed to place the content in close proximity to the client requesting it. CDN provides high content availability mostly because of its highly effective content distribution and replication strategy. The replication strategy minimizes content retrieval delay and user latency, optimizes the available bandwidth, and limits the amount of internet traffic [43]. Clustering technique has been implemented in content replication, which has already been discussed in the previous section. To simplify the replication, clustering can be leveraged to make groups of content based on their popularity in specific time and space [23, 44]. Deep learning is also used to determine the importance of any particular document for any specific site. Statistical learning can be applied to decide the replication degree of the internet content based on the popularity prediction. Video on demand predicts the replication videos by ranking them based on their hotness. Support vector machine (SVM) can be used to classify the Internet content as popular or non-popular. Non-popular content means their access pattern does not get changed with their request process. On the other hand, the access curve gets increased with time for popular content. Reinforcement learning is an area of machine learning that makes suitable sequences of decisions by interacting with the environment [44]. This reinforcement learning has been adopted in CDN to dynamically learn the optimal number of duplications for a particular content [45]. Even RL has been proposed as a choice for automating CDN parameters by Google, Microsoft, Facebook, etc. [46-48]. The effectiveness of CDN also depends on the cache organization. In case of a cache miss when a surrogate redirects the request, a crucial decision needs to be made by the server itself as to whether to admit the content and cache; also, which content to evict in the case of a full cache with no extra room. Cache organization is done with RL-Cache, which uses reinforcement learning (RL) to efficiently process requests and make admission decisions with minimum computational overhead [49].

Fuzzy CDN is an emerging concept in CDN as it helps to select a surrogate for a particular user request considering the critical issue called load balancing [50]. Unlike Boolean logic, fuzzy logic can make decisions based on intermediate values between true and false. Fuzzy logic can be used to implement adaptive routing algorithms as it considers the status information being exchanged between the surrogates. Different network parameters, such as queue size, service time and round-trip time (RTT), can be realized using the membership functions for the input. These input functions are transformed into output membership functions through fuzzy rules. The Wang-Mendel algorithm can be used to generate inconsistent fuzzy rules [51]. Fuzzification can also be done on other parameters like bandwidth, hard disk and connectivity, and the replica server with the highest fuzzy decision value is considered as the most suitable replica server to serve the client request [52, 53]. Fuzzy-based dynamic load balancing minimizes server load, network latency and packet loss for selecting an appropriate surrogate over the network [54].

6.4 Pandemic and CDN

In March of 2020, life came to a stop. That is when the COVID-19 pandemic started driving decisions on how we live, work, interact, and learn, i.e., our existence was completely driven by this virus.

In the month of December in the year 2020, the first official human COVID-19 infection was reported in Wuhan, China [51]. On March 11, 2020, the World Health Organization (WHO) declared that the COVID-19 outbreak was a pandemic. By April, 2020, almost 90 countries with more than half of the world's population imposed a lockdown to reduce community spreading. A single protocol was followed all over the world – social distancing – to bring down the death toll. Nations adopted two preventive measures – social distancing and quarantine – to restrict the movements of their populations by forcing them to stay at home. This isolation guideline led us to a virtual existence, both personally and professionally. Suddenly this unprecedented lockdown put a big question mark over our education, health, tourism, construction, aviation, transportation, agriculture, etc. In other words, the whole world economy was faced with the biggest challenge ever.

On March 24, 2020, the Government of India declared a nationwide lockdown which affected the entire education system the most. To counterattack the damage caused by the pandemic, educational institutes all over the world went to digital platforms as an alternative solution. Since then, digital education is becoming more popular and is being combined with traditional classroom teaching. This hybrid mode is gaining importance in the National Education Policy released by the Union Government. After the USA, India has risen to become the second largest market for Massive Open Online Courses (MOOCs).

For many, the world became the four walls of their homes. Due to the prolonged quarantine and lockdown, the entire global population is facing a severe social crisis. The prolonged stress due to social isolation has had a serious effect on mental health, resulting in psychological problems such as depression, frustration, and anxiety. Though this virus has affected every segment of the population, young adults and children are more prone to develop these anxiety symptoms because of the sudden change in their lifestyle. The spread of COVID-19 forced most of the companies worldwide to lock their offices, resulting in an increase of those working from home. Children were unable to participate in outdoor games and enjoy school life, and were confined to rooms. Young adults started facing the problem of loneliness as they became disconnected from their friends, family, and co-workers, which has had an effect on their socioemotional well-being. This entire scenario has led us towards virtual existence. Educational institutes all over the globe have come up with digitized education, and due to the confinement there has been a surge in watching TV and consumption of online streaming.

As the demand for the popular digital platforms Zoom, DropBox, and Netflix has gone up, due to lockdown-related supply chain issues, they needed to scale up their facilities for uninterrupted service through CDN. The CDN service providers have identified the changes in the traffic pattern during the recent lockdown: the traffic decreased on campuses and in office areas while it went up in residential areas [52]. The CDN is essential for handling the surge in demand for video streaming [53]. Netflix is a leading American over-the-top content platform that provides subscription-based streaming service [54]. Netflix has 209 million subscribers, including 72 million in the United States and Canada as of July 2021, which significantly increased during lockdown [55]. The key factor behind the success of Netflix is that it distributes most of its content through Open Connect CDN. Along with the risk to life, COVID-19 has also magnified the risk level of internet services because of their dependency on remote connectivity to a great extent. This external dependency provides more opportunities to the bad actors to monetize DDoS attacks, which will be discussed in detail in the next section.

6.5 Security Threats in CDN

Besides all the advantages provided by CDN, it also has some disadvantages. The primary concern for CDN is obviously the cost to be paid to the third-party network provider for server deployment and maintenance. Another matter of concern is loss of control due to the content provider having to hand over the control of their own website to a third party, which means providing all business-related information to the service provider. This particular aspect of CDN can result in a serious security breach.

Denial of Service (DoS) attacks, particularly Distributed Denial of Service (DDoS) attacks, are considered one of the major threats in today's internet [60]. A DDoS attack is a malicious attempt to disrupt the regular traffic of a targeted website by overwhelming requests from a botnet [61]. This attack results in huge economic loss as it has a direct impact on the victimized organization's reputation [62]. Any organization having an online presence can become the victim of this threat. Handling this security threat is very crucial in CDN, as CDN providers have the main aim of serving the customer with almost no delay. Therefore, it is not only important to protect the data but also ensure optimal CDN availability by navigating the malicious attacks [63]. Standard CDN is considered to protect their customers from layer 7 DDoS attack [64]. CDN can absorb this attack with almost negligible effect with the help of a number of servers and vast resources scattered all over the globe.

Another aspect of CDN security is to provide data protection. Edges in CDN cache the content to optimize the delay. But caching sensitive content to all the servers increases the probability of hacking. Instead, replicating only static and publicly available data minimizes the threat to a great extent [65]. Popularity of content results in information leakage, which leads to compromising users' privacy and revealing business-specific confidential information to untrusted CDN providers. Searchable encryption (SE) can be used to encrypt the objects and the request to the objects [66].

Blockchain is a decentralized ledger of records, called blocks, which are connected via cryptography. It was initially introduced for cryptocurrency known as bitcoin [67]. This emerging technology can be applied in CDN to provide security to the CDN customers. Blockchain-based CDN (B-CDN) provides a decentralized and secured platform for the users to connect with content providers [68-70]. The B-CDN is a network consisting of a chain of blocks which is responsible for validating and adding new blocks to the network and also authenticating both content providers and users. The users benefit from the security aspect as they are represented by virtual identity, which helps the user with a single registration when asking for services from different content providers.

6.6 Conclusion

In this chapter, we focused on the importance of emerging technology content delivery networks. The chapter identified the architectural components of CDN along with all major challenges involved in making decisions in all aspects. It was pointed out that the key areas of server placement, content placement, request routing and load balancing have attracted researchers for enhancing the efficiency of the entire system. Several server placement algorithms, content caching techniques, request routing approaches and load balancing schemes were analyzed along with their pros and cons. How CDN keeps on using cutting-edge technologies of AI and ML to optimize its response time with a disruption-free content delivery was also discussed. Next, there was a discussion of how CDN plays a very important role during pandemics by providing exhaustive support behind almost all the digital platforms and helps us in our virtual existence. Finally, we identified the drawback of CDN from a security perspective and how CDN can protect its customer and content from malicious attacks. From the current scenario and vast application domain it can be concluded that CDN is an unparalleled solution for today's digital world.

References

1. Usmanova, N., & Samandarov, B. (2020, November). Overlay Networking Issues: Implementation The Functional Components. In *2020 International Conference on Information Science and Communications Technologies (ICISCT)* (pp. 1-4). IEEE. DOI: 10.1109/ICISCT50599.2020.9351404.
2. Kim, J. Y., & Choi, J. K. (2014, October). Decentralized content delivery scheme using in-network caching. In *2014 International Conference on Information and Communication Technology Convergence (ICTC)* (pp. 901-902). IEEE. DOI: 10.1109/ICTC.2014.6983328.
3. Haribowo, Y., & Kistijantoro, A. I. (2012, April). Performance analysis of content-based mobile application on content delivery networks. In *2012 International Conference on Cloud Computing and Social Networking (ICCCSN)* (pp. 1-4). IEEE.
4. Sarkar, D., Rakesh, N., & Mishra, K. K. (2015, September). Problems in Replica Server Placement (RSP) over Content Delivery Networks (CDN). In *Proceedings of the Sixth International Conference on Computer and Communication Technology 2015* (pp. 175-179). DOI: 10.1145/2818567.2818600.
5. Elkotob, M., & Andersson, K. (2012, December). Challenges and opportunities in content distribution networks: A case study. In *2012 IEEE Globecom Workshops* (pp. 1021-1026). IEEE.
6. Stamos, K., Pallis, G., Vakali, A., & Dikaiakos, M. D. (2009, June). Evaluating the utility of content delivery networks. In *Proceedings of the 4th edition of the UPGRADE-CN workshop on Use of P2P, GRID and agents for the development of content networks* (pp. 11-20).
7. Rafetseder, A., Metzger, F., Stezenbach, D., & Tutschku, K. (2011, September). Exploring YouTube's content distribution network through distributed application-layer measurements: a first view. In *Proceedings of the 2011 International Workshop on Modeling, Analysis, and Control of Complex Networks* (pp. 31-36).
8. Chard, K., Caton, S., Rana, O., & Katz, D. S. (2012, November). A social content delivery network for scientific cooperation: Vision, design, and architecture. In *2012 SC Companion: High Performance Computing, Networking Storage and Analysis* (pp. 1058-1067). IEEE.
9. Gracia-Tinedo, R., S'nchez-Artigas, M., & García-López, P. (2012, June). FriendBox: A hybrid F2F personal storage application. In *2012 IEEE Fifth International Conference on Cloud Computing (CLOUD 2012)* (pp. 131-138). IEEE.
10. Vakali, A., & Pallis, G. (2003). Content delivery networks: Status and trends. *IEEE Internet Computing*, 7(6), 68-74.
11. Pathan, A. M. K., & Buyya, R. (2007). A taxonomy and survey of content delivery networks. *Grid Computing and Distributed Systems Laboratory, University of Melbourne, Technical Report*, 4(2007), 70.
12. Qiu, L., Padmanabhan, V. N., & Voelker, G. M. (2001, April). On the placement of web server replicas. In *Proceedings IEEE INFOCOM 2001. Conference on Computer Communications. Twentieth Annual Joint Conference of the IEEE Computer and Communications Society (Cat. No. 01CH37213)* (Vol. 3, pp. 1587-1596). IEEE.
13. Szymaniak, M., Pierre, G., & Van Steen, M. (2006). Latency-driven replica placement. *IPSJ Digital Courier*, 2, 561-572.
14. Jamin, S., Jin, C., Kurc, A. R., Raz, D., & Shavitt, Y. (2001, April). Constrained mirror placement on the Internet. In *Proceedings IEEE INFOCOM 2001. Conference on Computer Communications. Twentieth Annual Joint Conference of the IEEE Computer and Communications Society (Cat. No. 01CH37213)* (Vol. 1, pp. 31-40). IEEE.
15. Yang, M., & Fei, Z. (2003, May). A model for replica placement in content distribution networks for multimedia applications. In *IEEE International Conference on Communications, 2003. ICC'03.* (Vol. 1, pp. 557-561). IEEE.

16. Rodrigues, M., Moreira, A., Neves, M., Azevêdo, E., Sadok, D., Callado, A., & Souza, V. (2013, April). Optimizing cross traffic with an adaptive CDN replica placement strategy. In *Proceedings of the 46th Annual Simulation Symposium* (pp. 1-8).

17. Jafari, S. J., & Naji, H. (2013, May). GeoIP clustering: Solving replica server placement problem in content delivery networks by clustering users according to their physical locations. In *The 5th Conference on Information and Knowledge Technology* (pp. 502-507). IEEE.

18. Yin, H., Zhang, X., Zhan, T., Zhang, Y., Min, G., & Wu, D. O. (2013). NetClust: A framework for scalable and pareto-optimal media server placement. *IEEE Transactions on Multimedia*, 15(8), 2114-2124.

19. Sarkar, D., Rakesh, N., & Mishra, K. K. (2018). Population-based clustering to enhance the utilization of surrogate in Content Delivery Networks. *Intelligent Decision Technologies*, 12(4), 453-460.

20. Al-Shayeji, M. H., Rajesh, S., Alsarraf, M., & Alsuwaid, R. (2010, December). A comparative study on replica placement algorithms for content delivery networks. In *2010 Second International Conference on Advances in Computing, Control, and Telecommunication Technologies* (pp. 140-142). IEEE.

21. Sarkar, D., Rakesh, N., & Mishra, K. K. (2018). Broadcast Storm Problem—A Hidden Consequence of Content Distribution in Content Delivery Networks. In *Networking Communication and Data Knowledge Engineering* (pp. 155-165). Springer, Singapore. DOI: 10.1007/978-981-10-4585-1_13.

22. Markatos, E. P., & Chronaki, C. E. (1998, July). A top-10 approach to prefetching on the web. In Proceedings of INET (Vol. 98, pp. 276-290).

23. Chen, Y., Qiu, L., Chen, W., Nguyen, L., & Katz, R. H. (2003). Efficient and adaptive Web replication using content clustering. *IEEE Journal on Selected Areas in Communications*, 21(6), 979-994.

24. Wu, B., & Kshemkalyani, A. D. (2005). Objective-optimal algorithms for long-term Web prefetching. *IEEE Transactions on Computers*, 55(1), 2-17.

25. Fujita, N., Ishikawa, Y., Iwata, A., & Izmailov, R. (2004). Coarse-grain replica management strategies for dynamic replication of Web contents. *Computer Networks*, 45(1), 19-34.

26. Sivasubramanian, S., Szymaniak, M., Pierre, G., & Steen, M. V. (2004). Replication for web hosting systems. *ACM Computing Surveys (CSUR)*, 36(3), 291-334.

27. Wang, L., & Pai, V. (2002). The Effectiveness of Request Redirection on CDN Robustness. In *5th Symposium on Operating Systems Design and Implementation (OSDI 02)*. (pp. 345-360).

28. Szymaniak, M., Pierre, G., & van Steen M (2003). NetAirt: A Flexible Redirection System for Apache. In *Proceedings of International Conference WWW/Internet, Algrave, Portugal.*

29. Aggarwal, A., & Rabinovich, M. (1998). Performance of dynamic replication schemes for an internet hosting service. *Technical Report, HA6177000-981030-01-TM, AT&T Research Labs*, Florham Park, NJ, USA.

30. Delgadillo, K. (1997). "Cisco DistributedDirector," *Cisco White Paper, Cisco Systems, Inc.*.

31. Pierre, G., & Van Steen, M. (2006). Globule: a collaborative content delivery network. *IEEE Communications Magazine*, 44(8), 127-133.

32. Andrews, M., Shepherd, B., Srinivasan, A., Winkler, P., & Zane, F. (2002, June). Clustering and server selection using passive monitoring. In *Proceedings of Twenty-First Annual Joint Conference of the IEEE Computer and Communications Societies* (Vol. 3, pp. 1717-1725). IEEE.

33. Sharma, S., Singh, S., & Sharma, M. (2008). Performance analysis of load balancing algorithms. *World Academy of Science, Engineering and Technology*, 38(3), 269-272.

34. Eager, D. L., Lazowska, E. D., & Zahorjan, J. (1986). Adaptive load sharing in homogeneous distributed systems. *IEEE Transactions on Software Engineering*, (5), 662-675. DOI: 10.1109/TSE.1986.6312961.

35. Xu, Z., & Huang, R. (2009). Performance study of load balancing algorithms in distributed web server systems. *CS213 Parallel and Distributed Processing Project Report.*

36. McEntire, P. L. (1984). *Distributed Computing: Concepts and Implementations.* IEEE Press.

37. Malik, S. (2000). Dynamic load balancing in a network of workstations. *Pap. Parallel Process. Course, Carlet, (219762).*

38. Leinberger, W., Karypis, G., Kumar, V., & Biswas, R. (2000, May). Load balancing across near-homogeneous multi-resource servers. In *Proceedings 9th Heterogeneous Computing Workshop (HCW 2000)(Cat. No. PR00556)* (pp. 60-71). IEEE. DOI: 10-7695-0556-2/00, 2000 IEEE.

39. Maki, N., Shinkuma, R., Mori, T., Kamiyama, N., & Kawahara, R. (2013, April). A periodic combined-content distribution mechanism in peer-assisted content delivery networks. In *2013 Proceedings of ITU Kaleidoscope: Building Sustainable Communities* (pp. 1-8). IEEE.

40. Leong, D., Ho, T., & Cathey, R. (2009, March). Optimal content delivery with network coding. In *2009 43rd Annual Conference on Information Sciences and Systems* (pp. 414-419). IEEE.

41. Gupta, P., Goyal, M. K., & Gupta, N. (2015). Reliability aware load balancing algorithm for content delivery network. In *Emerging ICT for Bridging the Future-Proceedings of the 49th Annual Convention of the Computer Society of India (CSI) Volume 1* (pp. 427-434). Springer, Cham. DOI: 10.1007/978-3-319-13728-5_48

42. Abbas, O. A. (2008). Comparisons between data clustering algorithms. *International Arab Journal of Information Technology (IAJIT)*, 5(3), pp. 320-325.

43. Dukkipati, N., & McKeown, N. (2006). Why flow-completion time is the right metric for congestion control. *ACM SIGCOMM Computer Communication Review*, 36(1), 59-62.

44. Haj-Ali, A., Ahmed, N. K., Willke, T., Gonzalez, J., Asanovic, K., & Stoica, I. (2019). A view on deep reinforcement learning in system optimization. *arXiv preprint arXiv:1908.01275.*

45. Wang, Z., Du, S., & Ren, M. (2021). NCDN: A Node-Failure Resilient CDN Solution with Reinforcement Learning Optimization. Mobile Information Systems, 2021. DOI: 10.1155/2021/6663243.

46. Bychkovsky, V., Cipar, J., Wen, A., Hu, L., & Mohapatra, S. (2018). Spiral: Self-tuning services via real-time machine learning. Blog post here. Available at https://code.fb.com/data-infrastructure/spiral-self-tuning-services-via-real-time-machine-learning/, accessed 07/10/18.

47. Dean, J. (2018). Is google using reinforcement learning to improve caching. *Personal communication on*, 09-27.

48. Berger, D. S. (2018, November). Towards lightweight and robust machine learning for cdn caching. In *Proceedings of the 17th ACM Workshop on Hot Topics in Networks* (pp. 134-140).

49. Kirilin, V., Sundarrajan, A., Gorinsky, S., & Sitaraman, R. K. (2020). Rl-cache: Learning-based cache admission for content delivery. *IEEE Journal on Selected Areas in Communications*, 38(10), 2372-2385.

50. de Oliveira, T. Q., & Fernandez, M. P. (2013, March). FuzzyCDN: fuzzy redirection algorithm. In *2013 IEEE 27th International Conference on Advanced Information Networking and Applications (AINA)* (pp. 437-444). IEEE. DOI: 10.1109/AINA.2013.112

51. Wang, L. X., & Mendel, J. M. (1992). Generating fuzzy rules by learning from examples. *IEEE Transactions on Systems, Man, and Cybernetics*, 22(6), 1414-1427.

52. Chen, J. B., & Liao, S. J. (2010, September). A fuzzy-based decision approach for supporting multimedia content request routing in cdn. In *International Symposium on Parallel and Distributed Processing with Applications* (pp. 46-51). IEEE.

53. Cai, L., Ye, J., Pan, J., Shen, X. S., Mark, J. W. (2006). Dynamic server selection using fuzzy inference in content distribution networks. *Computer Communications*, 29(8), 1026-1038. DOI: 10.1016/j.comcom.2005.06.001.

54. Roy, S., Bose, R., & Sarddar, D. (2015, June). Fuzzy based dynamic load balancing scheme for efficient edge server selection in Cloud-oriented content delivery network using Voronoi diagram. In *2015 IEEE international advance computing conference (IACC)* (pp. 828-833). IEEE. DOI: 10.13140/RG.2.1.1996.5287.

55. WHO, C. O. (2020). World health organization. Responding to Community Spread of COVID-19. Reference WHO/COVID-19/Community_Transmission/2020.1.

56. How Zoom, Netflix, and Dropbox are Staying Online During the Pandemic (datacenterknowledge.com)

57. 5 CDN Metrics You Should Care About. How a CDN Can Help You Meet Your Video Streaming Demands - Intequus. https://www.intequus.com/cdn-metrics/

58. https://www.netflix.com/

59. Kastrenakes, Jacob (April 20, 2021). Netflix subscriber growth is stalling as it runs low on hits. *The Verge*.

60. Poongothai, M., & Sathyakala, M. (2012, December). Simulation and analysis of DDoS attacks. In *2012 International Conference on Emerging Trends in Science, Engineering and Technology (INCOSET)* (pp. 78-85). IEEE. DOI: 10.1109/INCOSET.2012.6513885.

61. https://www.cloudflare.com/en-in/learning/ddos/what-is-a-ddos-attack/

62. https://www.globaldots.com/resources/blog/content-delivery-network-explained/#Content_Delivery_Networks_and_Security

63. https://www.akamai.com/uk/en/resources/cdn-security.jsp

64. Triukose, S., Al-Qudah, Z., & Rabinovich, M. (2009, September). Content delivery networks: protection or threat?. In *European Symposium on Research in Computer Security* (pp. 371-389). Springer, Berlin, Heidelberg. DOI: 10.1007/978-3-642-04444-1_23.

65. https://www.teridion.com/cdns-safe-cdn-security/

66. Cui, S., Asghar, M. R., & Russello, G. (2017, October). Privacy-preserving content delivery networks. In *2017 IEEE 42nd Conference on Local Computer Networks (LCN)* (pp. 607-610). IEEE. DOI: 10.1109/LCN.2017.27.

67. Wright, C. S. (2008). Bitcoin: a peer-to-peer electronic cash system. *Available at SSRN 3440802*.

68. Seth, B., Dalal, S., Le, D. N., Jaglan, V., Dahiya, N., Agrawal, A., ... & Verma, K. D. (2021). Secure Cloud Data Storage System Using Hybrid Paillier–Blowfish Algorithm. *CMC-Computers, Materials & Continua*, 67(1), 779-798.

69. Pramanik, S., Ghonge, M. M., Mangrulkar, R., & Le, D. N. (Eds.). (2022). *Cyber Security and Digital Forensics: Challenges and Future Trends*. John Wiley & Sons.

70. Vu, T. X., Chatzinotas, S., & Ottersten, B. (2019, April). Blockchain-based content delivery networks: Content transparency meets user privacy. In *2019 IEEE Wireless Communications and Networking Conference (WCNC)* (pp. 1-6). IEEE. DOI: 10.1109/WCNC.2019.8885904.

7

Challenges and Opportunities in Smart City Network Management Through Blockchain and Cloud Computing

Jayden Pires[1], Vinod Kumar Shukla[1], Leena Wanganoo[2], Sonali Vyas[3]

[1] Department of Engineering and Architecture Amity University, Dubai, UAE
[2] University of Petroleum and Energy Studies, Dehradun, India
[3] School of Computer Application, University of Petroleum and Energy Studies, India
Email: jaydenP@amitydubai.ae, vshukla@amityuniversity.ae, leenawanganoo@yahoo.com, svyas@ddn.upes.ac.in

Abstract

With the idea of new developing technologies and the rapid change in our technical environment, there is a need for smarter, more efficient technology to keep up with the needs of today's world. Thus, the idea of smart cities has come into being. The purpose of this chapter is to promote the idea of building a smart city and providing a framework for future development of the same. We are moving towards a more futuristic world where everything is connected and done over the internet, which will create more network dependency.

The problem with the concept of smart cities is the need for better technologies to improve the system of data transfer to address the problem of trust in the system, as massive amounts of people's private data is at stake with bad actors trying to hack the system to gain access to it. There is also the problem of the massive amounts of data produced in smart cities, so there is a need to provide a framework to efficiently store all the data.

These problems can be solved with the help of technologies such as blockchain and cloud computing. Cloud computing can be seen as a way of sharing resources offering compute, storage, databases, and network services that can be used on a larger scale. Cloud computing stores massive amounts of data, it can also be used to share and transfer data between networks in a cloud, it is cost-effective, and you can also retrieve lost data or backup data when needed. Despite cloud computing being so great, it also has its own drawbacks. The problem of security in cloud storage is ever-increasing; since cloud storage, centralized systems have been a major target for hackers and data breaches, and the data in the cloud can be tampered with and accessed by unauthorized users. If data is being transferred through the cloud it can be intercepted during transmission because of the problem of encryption. These problems can be solved as well as the system being improved and made faster with the help of blockchain.

Blockchain is a database or ledger that is shared across a network. The ledger is encrypted so that only people who are authorized can access the data. Blockchain will decentralize the data stored, which will improve the security of the data immensely. If a hacker targets a single block in a node, it will not have much effect and will continue to function. Block chain works by breaking the data into smaller chunks, and the chunks are distributed in a way that even if the network is down the data will still be available. While a file is being shared it is broken into smaller pieces and sent in a way that the file transfer is quicker. The files are encrypted using a cryptographic hash table and each individual shard or part of the file is also encrypted, making the security even tighter. With the help of blockchain, the storage of files will become much cheaper and more secure. A framework is provided to manage and secure the data acquired from IoT devices through encryption using Advanced Encryption Standard (AES) algorithm and storing the data in a private or public cloud depending on the data; and then decrypting the data and sending it to the users who requested the data or for further analysis. This chapter provides a framework for a smart city storage and security model and can be used as a prototype model for future development in the field.

Keywords: Smart city, blockchain, cloud computing, advanced encryption standard, network management, security

7.1 Introduction

The world is changing and the future is now. We are moving into a more futuristic world in which there are newer and more innovative designs that are created to improve our daily lifestyle and the efficiency of work that we do with the help of machines. In the futuristic world there will be connections everywhere, devices connected the majority of the time, machines cleaning the cities and many more advancements that help us achieve the goal of a smart city. Smart cities are the future, and as we implement more and more technologies into our lives there is a need for better security as well as a place to store all of the massive amounts of data that is being produced every day. This chapter discusses network management, blockchain, and cloud computing and their uses in a smart city. It also aims to implement the concept of blockchain for security, which will help secure the massive amounts of data produced in the smart cities. Cloud computing will help solve the problem of storing the data and provide the required/necessary space requirements needed for a smart city to increase its overall effectiveness and efficiency.

7.2 Literature Review

This literature review is divided in three sections from a network management point of view. Section 7.2.1 talks about smart city and various issues like network management and data storage. Section 7.2.2 talks about blockchain technology and its challenges in network management and other issues like smart contract, smart healthcare, and security frameworks. Section 7.2.3 is all about cloud computing and security challenges, and big data. From a smart city point of view, network management is a challenge and this has been covered in the following three sections of this literature review.

7.2.1 Smart City and Network Management

7.2.1.1 IoT Uses and Applications

A smart city focuses on using the new generation of technologies that are being made for all parts of our lives. With the help of sensors and other IoT devices, we can improve the quality of hospitals, railways, power grids, bridges, roads, tunnels, buildings, water systems and much more throughout the world with the connection of IoT devices over a network. Smart cities can integrate, process, and store the data of IoT with cloud computing and supercomputers [1].

What does a smart city do? A "smart city" is defined by IBM as the use of information and communication technology to sense, analyze, and integrate core systems' key information in running cities. At the same time, a smart city can intelligently respond to different kinds of needs, including daily work, environmental protection, public safety and city services, industrial and commercial activities [2].

Following are some of the steps being taken towards the goal of a smart city using the examples of Singapore and Dubai:

1. In Singapore, the government aims to make improvements to the economy by collecting data digitally through sensors linked with aggregation boxes. Tools are also being developed to create a dynamic 3D city model as well as a joint data platform, to simulate a smart city environment.

2. In Dubai, there is a plan to digitalize all government services, like communications, electricity, and economic benefits. A lot of the services provided by the government are already digitalized and accessible in an app called DubaiNow. There are already police stations where you can pay fines or report cases without needing to talk to a person.

7.2.1.2 IoT Infrastructure for Smart Cities

The internet of things (IoT) refers to interconnected objects, sensors, software, and many other technologies connected in a network that are able to transfer and manage data without the need for human effort. For a smart city to work properly it uses all types of different IoT devices. Jin *et al.* [3] have mentioned that IoT is the key enabler for smart cities, which can be classified into three domains (Figure 7.1):

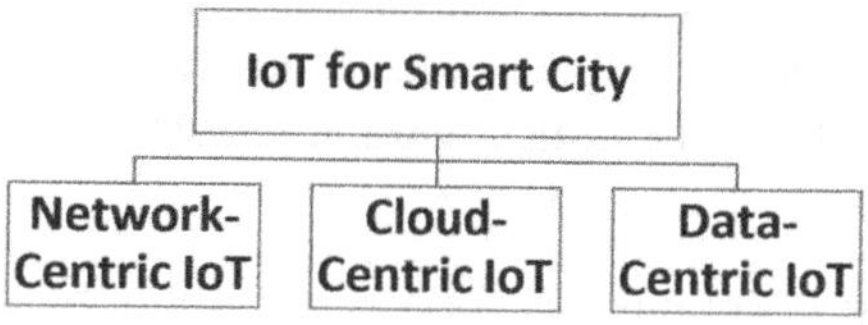

Figure 7.1: IoT for smart city.

- Network-Centric IoT: The concept of IoT can be looked at in two ways: "Internet" based and "Object" based. The internet-based architecture will be more of the services used through the internet while the data is contributed by the objects [3]. In the object-based architecture [4], the smart objects will take the main control.

  1. Sensing paradigm: There are mainly three main sensing paradigms (Figure 7.2): RFID, WSN and crowd sourcing.

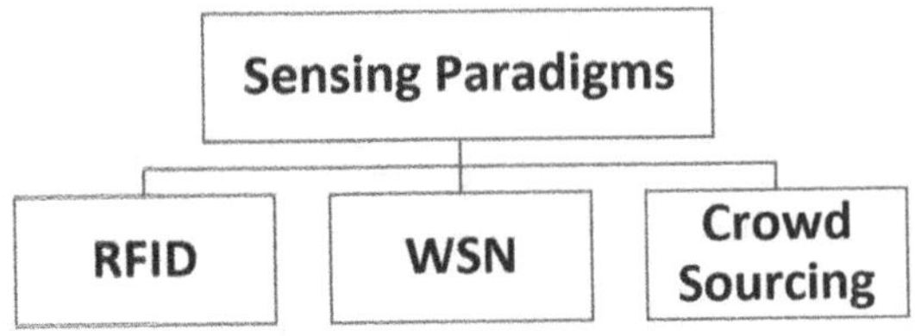

Figure 7.2: Different sensing paradigms.

- Radio-frequency identification (RFID): This technology can automatically identify the object to which the tags are connected by receiver via radio waves. RFID works in a similar way as barcoding but in this case, RFID does not have to be in the line-of-sight of the device, whereas barcode scanners need optical scanners. RFID is used in many industries like inventory management, ID badging, controlling access of restricted areas and many more [5].
- Wireless sensor networks (WSN): These allow the collection, analysis, processing and dissemination of information that is around in different types of environments. WSN plays an important role in urban sensing applications [3].
- Crowd sourcing: This is another paradigm which is also known as participatory sensing. It plays a major role in the interaction between government and citizens through advancements in smartphone technology [3].

2. Addressing scheme: An object can be identified uniquely, which is of major importance for the success and growth of IoT. This will allow the many devices that exist in a smart city to be controlled remotely

- Cloud-Centric IoT: To be able to integrate the ubiquitous urban sensing and smart city applications, and to show the potential of cloud computing, the sensing service providers can join the network and share their data in the cloud, analytical tool developers can provide their software tools, and computational intelligence experts can provide their machine learning and data mining tools to convert the data into useful knowledge. Cloud computing offers all these services as software or platforms, and the cloud is able to integrate all the needs of ubiquitous computing by providing the needed storage which can be scaled and the resources for a new business model as well.

- Data-Centric IoT: It is not unusual for a smart city to hold massive amounts of data when main devices in an IoT network are connected. Data-centric IoT makes sure that all areas of data flow, such as collection, storage and processing, and visualization of data is possible.

1. Data collection: A generalized framework for the collection of data is needed to effectively exploit spatial and temporal characteristics of data [3], both the data collected by sensors as well as data collected from mobile sensing infrastructures. Other than common sensors, such as RFIDs and WSN, participatory sensing is another upcoming sensing paradigm, where people instead of sensors take on the role of collecting the data as well as sharing of the sensory data [6]. Since it uses

people rather than sensors this will reduce the cost of buying expensive sensors to measure the environmental data, but it is difficult to judge the quality of the data acquired by the people. So, there is a need to certify privacy as well as the trust of data collected [3].

2. Data management and processing: Acquiring information that is meaningful from the raw data that is collected is very important. It usually involves pre-processing of the data and event detection [3]. Events can be detected in time frames. For a smart city, when an event happens, it is necessary to know how the algorithms work to check and compare data on a large scale of space and time. Then, to take this data and further make the data into something we can use as information, we need to use computational intelligence techniques such as genetic algorithms, neural networks and evolutionary algorithms [7]. These techniques will help make decisions which are automated, but due to the huge amount of data that is being acquired and stored, the expiry and ownership of data is a problem. So, there needs to be an energy-effective solution and a way to intelligently sort the data.
3. Data interpretation: The information of data in smart city applications needs to be put in a form that can be interpreted by the user of the data. There are newer and newer technologies that are being used to collect data and present it in a form that can be visualized. An example of this is when we switched from CRT displays to Plasma, LED and AMOLED displays which have allowed us to view the data and better navigate the data.

7.2.1.3 Security Problems in Smart Cities

Due to the nature of resources of the devices used in a smart city, the smart city is open to many security attacks. It is important to find out about these threats and what their consequences might be and create an effective solution to them. A lot of research has been conducted in this field, such as the Open Web Application Security Project (OWASP) enlisting common security attacks, Computer Emergency Response Teams (CERT) providing graphical representation of potential vulnerabilities, and G Cloud presenting a series of Cloud Computer Service Provider (CCSP) requirements [1, 8, 9]. (G Cloud is a framework for cloud-based solutions that are put forward by the cloud service sellers, which is also owned by Google.) The common threats can be classified as follows [10]:

- Threats on Availability: Threats related to the resources being used by unauthorized users.
- Threats on Integrity: Threats related to the change to data that was not authorized by either manipulation or corrupted information.
- Threats on Confidentiality: Threats related to the disclosure of the data held or sensitive information by the organizations to an unauthorized entity.
- Threats on Authenticity: Threats related to users that are unauthorized gaining access to the resources and sensitive information in those resources.
- Threats on Accountability: These include denial of transmission or reception of a message by the corresponding entity.

7.2.1.4 Data Storage Problems in Smart City

The main reasons for creating smart cities is to improve sustainability, to create effective and efficient economic development, and to improve the overall quality of life. However, these improvements do not come at a cheap price nor do they come in a perfect manner. There will be many problems that we have to deal with, especially the problem of storage of the massive amounts of data that the smart cities produce. Cities in the past have used video surveillance as a tool to improve public safety, but a smart city will use these videos for many different purposes, like traffic management, smart energy usage using heat sensors and much more [11].

During these times of the internet of things, artificial intelligence, and virtual reality, newer and better sensors are being developed as time goes on and there is a need to store these massive amounts of data somewhere. These large volumes of data are known as big data. Big data that is being received through the many devices needs a place in which the data can be stored. There is also the problem of security of the data. This data needs to be kept in a manner that is secure but at the same time can be shared throughout the city whenever necessary. Smart cities have the ability to improve the workings of many areas in almost every field of work and improve the quality of life; hence, it is necessary to find a good storage solution to ensure that the workings of the smart cities go smoothly.

7.2.2 Blockchain and Network

7.2.2.1 Blockchain Technology

Blockchain is a peer-to-peer distributed ledger technology which records transactions, agreements, contracts, and sales [12]. It was originally made in conjunction with cryptocurrency to support it, and can be used for any transactions without the need for there to be a middleman. The biggest advantage of blockchain is its decentralized system; any attack has to compromise at least 51% of the systems to get past the hashing power of the targeted network. So, we can say that it is very difficult to launch an attack against a blockchain network [10].

Example: There are two entities, A and B, and the type of blockchain system we will be using is for a parking system where A is paying a fee for parking to the landlord B. This transaction is shown online as a block that includes information such as block numbers, previous blocks, transaction records and proof of work; this block is shown to every entity in the network. The other entities make sure that more than 50% of the entities approve the block, and then and only then will the transaction be confirmed and go through. Then the fee for the parking is transferred from A's account to landlord B's account [10].

7.2.2.2 Smart Contracts

There have been many blockchain platforms written in different computer languages which have a certain set of rules; this is called a smart contract [13]. Smart contracts are self-executing contracts because the terms of the contract between the buyer and the seller are written in the lines of code. The agreement and the code are stored in a decentralized, distributed, blockchain network. The code takes control over the execution and transactions, which are irreversible and trackable. Smart contracts allow secure and trusted agreements and transactions to be carried out without the need for a legal system, middleman, central authority, or any external mechanism [14].

7.2.2.3 Security Framework

The security framework proposed by Biswas and Muthukkumarasamy [10] (Figure 7.3) highlights how blockchains work with smart city infrastructure and how data travels from the physical layer all the way to the applications.

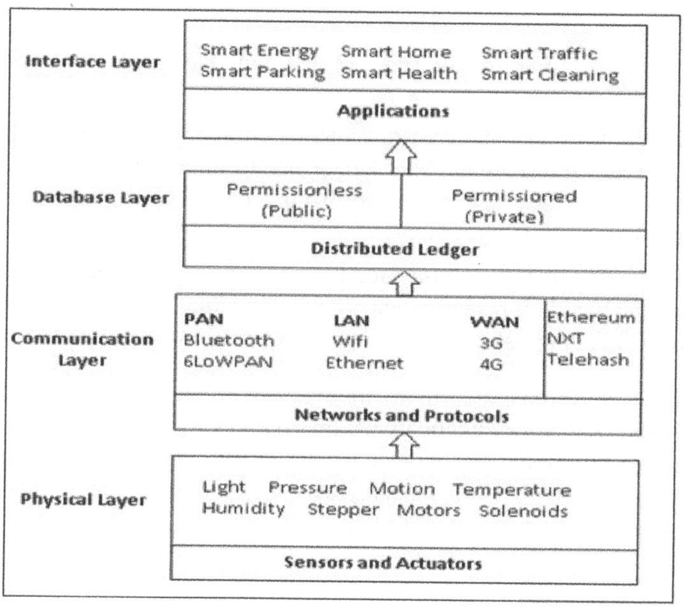

Figure 7.3: Smart city security framework proposed by Biswas and Muthukkumarasamy [10].

- Physical Layer: In the figure shown above, smart city devices are equipped with sensors and actuators which collect data and transfers it to the upper layer protocols. Devices such as the Fitbit Acer and Nest thermostat are open to security attacks because of light security care and the access control mechanism [15]. The data that is produced in smart devices does not have a single standard to follow, the vendors are required to have agreed-upon implementation and communication standards to overcome problems in smart devices as well as to implement cross-functionality [10].

- Communication Layer: In this layer, smart city networks use different communication mechanisms, such as Bluetooth, Wi-Fi, 3G, and 4G, to exchange information between systems. This is where the blockchain protocols need to be implemented to provide security and privacy of the data that is transferred. The use of multiple blockchains and blockchain access layers to provide application-specific functionalities to fix blockchain's problem with existing communication protocols due to the application varies from application to application [10].

- Database Layer: In blockchain, a distributed ledger is a type of decentralized database that stores records one after another. Each record in the ledger includes a time stamp and a unique cryptographic signature. The transactions and the history of the ledger are verifiable and auditable by any authorized users. There are two different types of distributed ledgers (Figure 7.4) that are used [10]:

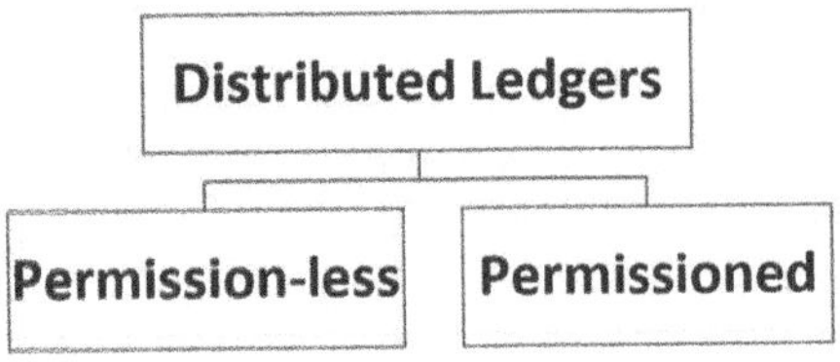

Figure 7.4: Types of distributed ledgers.

- Permissionless: The main reason to use a permissionless ledger is that it is censorship-resistant and transparent. This ledger has to maintain complex records that are shared, and it takes more time to reach a consensus. They are more prone to security attacks [10].
- Permissioned: It is recommended to use permissioned ledgers because they are more scalable, their performance is better, and they have better security, especially for real-time applications like traffic systems in a smart city [10].

- Interface Layer: In this layer there are many smart applications that work with each other to make decisions effectively. A smartphone application can allow you to connect to your smart home and activate the air conditioner minutes before you enter the home from a remote location. The applications need to be integrated properly; if not, the application is prone to attacks [10].

7.3 Blockchain and Smart City

Blockchain can contribute a lot to smart cities. Some sectors in which blockchain can help improve how a smart city works are given below.

7.3.1 Smart Healthcare

Healthcare centers carry huge amounts of data from patients that is transferred from healthcare providers to insurance companies. Since this data has a lot of private information about the patient, there is a need for a high level of security and control over who can access the data. With the help of blockchain technologies we can ensure the data integrity and the swift exchange of the patient's medical records. This also helps with adjudication of insurance claims, as well as provides a high level of security and reliability, transparency, and shared access to authorized users [11-16].

7.3.2 Smart E-Voting

There is a need for the government to move to an online system and adapt e-voting for elections. E-voting provides convenience, accountability, and easier access to democratic elections. In recent months, COVID-19 has also shed more light on the idea of online voting, especially for public safety and a smoother democratic process. When used in conjunction with blockchain, e-voting provides high authentication capabilities that allow it to store the votes securely and offers more transparency to elections. With blockchain, the cost of the election process will be lower, the election process will be more transparent, and there will be less manipulation of voting data [16].

7.3.3 Smart Logistics and Supply Chains

Blockchain can improve the performance of logistics and supply chain operations in smart cities. It can simplify the logistics and supply chain processes of communication and information exchange between the parties involved [17]. Blockchain can help the stakeholders in the supply chain to efficiently manage the flow of goods and servers in the network. For logistics, block chain contributes a solution for the coordination of documents, reduction of processing times and approvals. Blockchain's features improve the competitive edge that supply chain industries have, reduce the costs of logistics, and reduce traffic congestion within the city [16].

7.4 Cloud Computing and Challenges

7.4.1 Cloud Computing

Cloud computing is a new style of computing in which the size can be scaled and the resources can be provided virtually though the internet. Cloud computing has become a necessity in today's world and a booming technology trend. It is beginning to reshape a lot of information technology processes and the IT market itself. Cloud computing technology uses various types of devices like personal computers, smartphones, laptops and personal digital assistants to access programs, provide storage for massive amounts of data, and a place to develop applications on a platform through the internet. There are many reasons why more and more companies or organizations are starting to pick up cloud computing services, some of which are the costs are low, services are readily available, and it is easy to scale [18].

7.4.1.1 Architectural Components

Cloud service models are usually shown by a given cloud infrastructure [19], represented in Figure 7.5. The architectural components of cloud service models are:

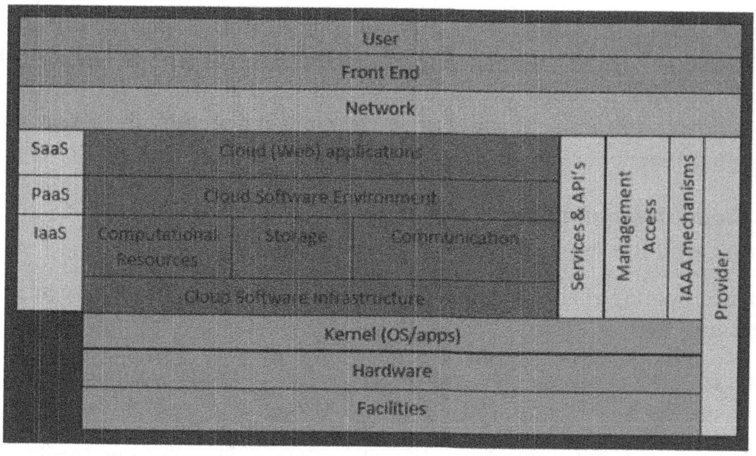

Figure 7.5: Architectural components of cloud service.

- User and Front End: Service Customer
- Kernel, Hardware, Facilities and Provider: Support IT Infrastructure

- Cloud's Services, Management Access, Mechanisms: Cloud-Specific Infrastructure

They are divided into:

- Software as a Service (SaaS): Cloud consumers put out their applications in an environment that allows them to host the service, which can be accessed through various clients like web browser, PDAs, and many more by the application users. The users of the cloud have no control over the cloud infrastructure. The cloud infrastructure uses a multi-tenancy system architecture, which organizes a signal logical environment in the SaaS cloud which allows optimization in terms of speed, disaster recovery, maintenance, and availability. Examples of SaaS are Google Mail and Google Docs [19].
- Platform as a Service (PaaS): PaaS cloud provides cloud components and also can be used for applications. Developers can build in a framework that is provided by PaaS, and it can be used to build and customize applications. Networking and server-side storage can be managed by a third-party enterprise and developers can focus on the management of the applications. The platform can be used over the web, and the developers will not have to worry about the operating systems, storage, updates for the software, or the infrastructure. They can freely design and create their software. This application can be scalable and can be shared throughout the cloud, and it is sometimes called middleware. An example is Google AppEngine [20].
- Infrastructure as a Service (IaaS): IaaS delivers cloud computing infrastructure through virtualization technology which includes servers, storage, network, and operating systems. It is very easy to scale and the use of resources can be automated by computing. It is available on-demand as a resource from third-party organizations, reducing the cost of components so physical maintenance will not be needed. The cloud servers are given to organizations through Application Programming Interfaces (API) or dashboards. IaaS clients have complete control over the entire infrastructure and can remotely access their servers and storage through a virtual data center in the cloud. An example is Amazon's EC2 [21].
- Data as a Service (DaaS): DaaS provides the clients with data storage services, a virtualized storage supply in a single cloud server. All data files are available to the client through a network usually over the internet. It uses a cloud-based technology that supports web services and service-oriented architecture; the data can be accessed through different devices. Examples are Amazon S3 and Google Bigtable [22].

7.4.1.2 Cloud Computing Applications

Following are a few applications of cloud computing [23]:

- Cloud computing enables organizations to procure dependable and secure data storage centers.
- Cloud computing allows data and resources like different equipment to be made easier.
- Over the internet, the cloud provides the client with numerous services and nearly infinite possibilities.
- Cloud computing does not require purchasing high-end equipment and the services of a cloud could be purchased from a third-party organization at a lower cost and are easier to use.

7.4.1.3 *Six Computing Paradigms*

These six computing paradigms (Figure 7.6) refer to how computers have evolved over six different phases, starting with dummy terminals all the way to grid computing and now cloud computing [24].

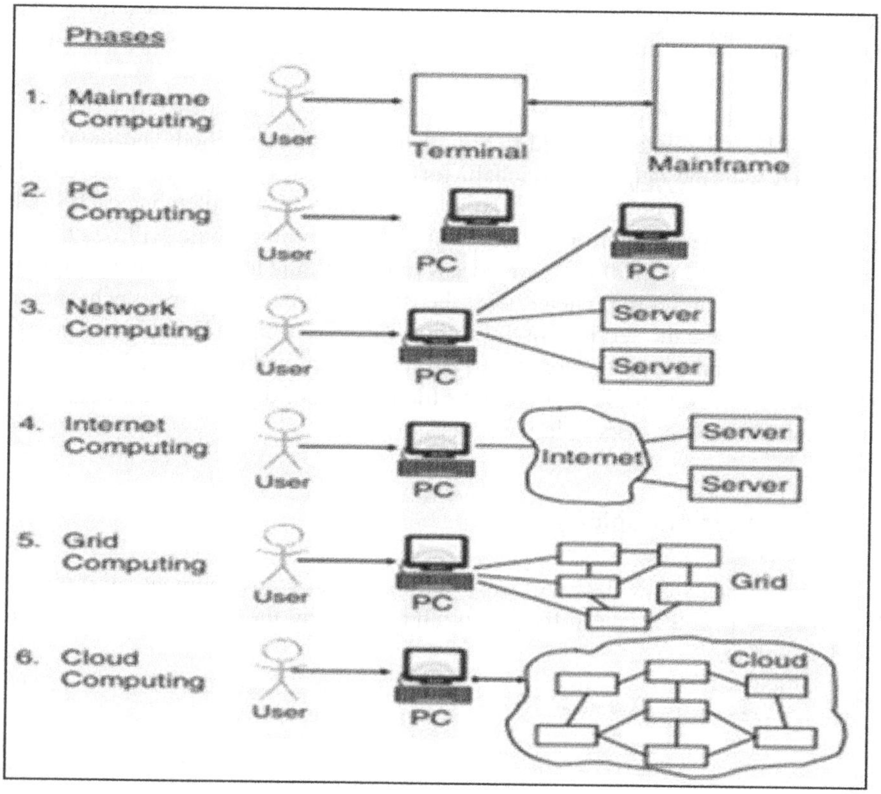

Figure 7.6: Six computing paradigms: From mainframe computing to internet computing, to grid computing and cloud computing [24].

The above figure proposes six phases of computing paradigms; in phase 1, many users shared powerful mainframes using dummy terminals [18]. In phase 2, PCs could work by themselves as long as they were connected to a network and met most of the requirements of the users. In phase 3, servers, laptops and PCs were connected in a local network so that resources could be shared between the devices and the performance could be increased. In phase 4, a local area network could be connected to another local area network which would form a global network like the internet, which would enable remote use of the applications and resources. In phase 5, through a shared computing system, grid computing allowed the sharing of computer power. In phase 6, cloud computer further enables the resources to be shared in a simpler and more scalable way. It may look as though the mainframe computing and cloud computing are similar through these designs, but they are different; mainframe computing provides finite computing power while in cloud computing there is almost infinite power, and the storage can be scalable. The dummy terminals in the mainframe computing act as a user interface, in cloud computing powerful PCs can provide the necessary computing power and support for crashes.

7.4.1.4 Big Data, Cloud Services, and Blockchain

We live in an age of big data, where data is being collected from individuals and organizations from all over the world on a very big scale, especially with IoT entering the picture [25]. In modern society, data is everywhere; every second or even millisecond around the world data is being transferred from one place to another [26]. There is great of demand for a way to securely store big data with lower costs and higher efficiency [27]. The entry of cloud computing on the scene has helped with the storage of the massive amounts of data produced by the numerous IoT devices [28].

Liu *et al.* [29] suggested the idea of a blockchain-based Data Integrity Service (DIS) framework where the data collected in cloud storage can be verified, and the data can be decentralized with the help of blockchain for better security because of the cloud being able to store massive amounts of data so securely, They also implemented DIS, which is a smart contract where the information received through blockchain will be encrypted so that unauthorized people will not have access to it. With the help of a smart contract, both parties will be able to communicate with the data after the nodes are synchronized [29]. No one can terminate the service but the author. The cloud-based IMS [30], in comparison to the blockchain proposed by Liu *et al.* [29], is efficient and reliable. (IMS is known for IP Multimedia Subsystem, basically allows the transfer of data over wireless connections and makes use of cloud services in the process.)

7.5 Research Methodology

Figure 7.7 represents how data will travel in a smart city environment, starting from the IoT devices, storing the data in the cloud, and retrieving the data as needed through the network or server.

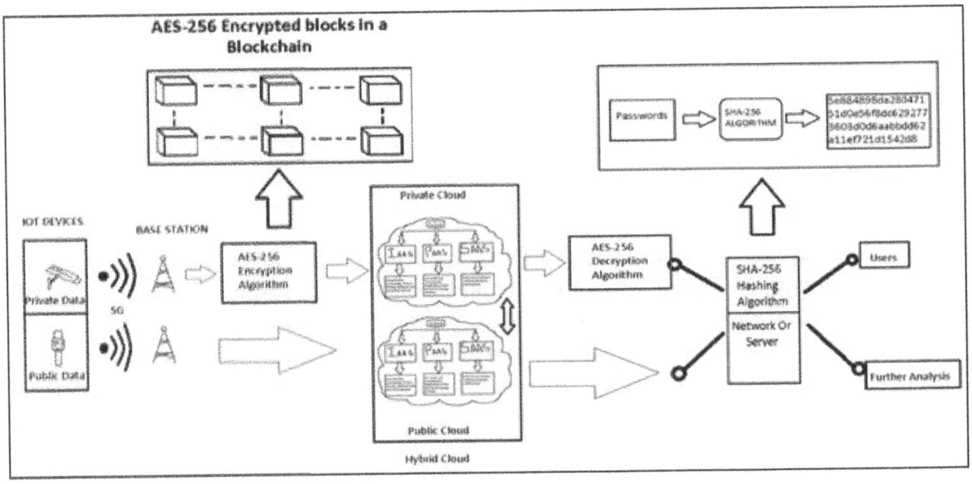

Figure 7.7: Framework for cloud computing, blockchain, and network management in smart city.

1. IoT Devices: Here, the IoT devices that collect data are put into two categories:

    - Private Data: IoT devices in the private data category collect data which will not be available for viewing by the general public. To depict private data I have taken the example of a smart camera with a motion sensor. Smart cameras with motion sensors are put in important places around the city, which enables us to control the lights in that area though motion detection as well as to keep areas of the city around banks, government buildings, etc., safe. The data that is provided in this scenario, like security footage, cannot be shared with the general public and thus will be stored in a private cloud.

- Public Data: IoT devices in the public data category collect data that can be viewed and used by the general public. To depict public data I have taken the example of a smart watch, which shows us data on the temperature and climate; and a GPS in which all data are available to the general public. The data that that is provided here will be stored in a public cloud.

2. Base Station or Routers: The data received from the IoT devices will be delivered into a cloud for sorting of the data, the base station will receive data from the IoT devices and transmit them to the sorting cloud. The transfer of data is done wirelessly through the internet via a 5G connection for increased speed and efficiency of data transfers.

3. Advanced Encryption Standard AES-256: The AES-256 algorithm is used here so that the data received by the private IoT devices can be encrypted and stored in the private cloud. AES is a symmetric block cipher; the same key is used for encryption and decryption. I have gone with the AES-256 algorithm rather than AES-128 because:

   a. In brute-force attacks, AES-256 is harder to crack than AES-128;
   b. The key for AES-256 is larger, which is more secure;
   c. Being bigger will require more computational power, but the extra security is worth it.

   Here, the data which is obtained is broken down into blocks. The blocks are then encrypted using an AES-256 encryption algorithm, which will improve the security greatly.

4. Hybrid Cloud: The hybrid cloud consists of two clouds, the public and the private cloud. A hybrid cloud refers to an environment made up of mixed storage, computing, and services. The public cloud can be used on the premises and the private clouds in a data center. Data that is collected does not always belong in the public cloud, for that reason I have taken a hybrid cloud [31].

   - Private Cloud: In the private cloud, only the private data is stored. The data stored here is encrypted. If needed, the private cloud can access the data in the public cloud, allowing the resources and data to be shared between them.
   - Public Cloud: In the public cloud, public data is stored. The data here is open data and can be used and viewed by all. The public cloud can request data from the private cloud only when it is requested and authorized.

5. AES Decryption Algorithm: The key to the earlier AES encrypted algorithm is stored here. This algorithm will decrypt the data that is received from the private cloud, so it is available for viewing by a request from authorized users.

6. Network or Server: The network or server is locked and is secured using a hashing algorithm. In the hashing algorithm SHA256, SHA stands for Secure Hashing Algorithm and 256 is the number of bits. This is a one-way algorithm which is mostly

used for passwords. Network or data center owners can secure their data with a SHA algorithm to provide extra security. Hashing of the data can also be used to check if the data has been modified or changed by comparing the data to the original hash to see if the data is intact.

In this framework we recommend using the SHA-256 algorithm rather than other SHA algorithms because:

1) The only extra security provided by SHA-512 is collision resistance for the algorithm, other than that, the SHA-256 hashing has the same infrastructure in which the extra cost is not worth it.

2) SHA-256 algorithm is faster than SHA-0, SHA-1, SHA-3 algorithms as of now [32].

3) SHA-256 has good security and uses the required bandwidth in a more efficient way [32].

7. Users or Further Analysis: If the data which is requested is private data then there will be a prompt that you will have to enter a password, while if public data is being requested it will be freely transferred to the user [32-35].

7.6 Conclusion

This chapter showed that the key problems of a smart city are the storage and security of data provided by the IoT devices. The chapter proposed a framework to solve these problems, with the help of cloud computing for data storage and blockchain for data security.

The provided framework would contribute a solution for the problem of security and storage in a smart city, but does not address some other limitations that are present, some of which are:

- Cost incurred to fully utilize the complete model is a lot.
- There is an issue of trust between the general public and the government, the people may or may not feel comfortable with the government having access to their data.
- If a private organization undertakes this model the government may restrict the working or operation of this infrastructure.
- There is a need for more educated personnel in blockchain, cloud computing and smart cities to maintain the working of this model.

This framework can be used as a base model for future developments in the field of smart city and IoT devices. With the invention of better sensors, security models and better ways to store data the creation of a smarter more connected world does not seem so far away. The aforementioned limitations can be countered with more resilient security models to develop the trust of people concerning their data, and the private and public sectors working in unison for further investments towards creating a fully operational smart city.

References

1. Su, K., Li, J., & Fu, H. (2011, September). Smart city and the applications. In *2011 International Conference on Electronics, Communications and Control (ICECC)* (pp. 1028-1031). IEEE.
2. Qin, H., Li, H., & Zhao, X. (2010). Development status of domestic and foreign smart city. *Global Presence*, 9, 50-52.
3. Jin, J., Gubbi, J., Marusic, S., & Palaniswami, M. (2014). An information framework for creating a smart city through internet of things. *IEEE Internet of Things journal*, 1(2), 112-121.
4. Lopez TS, R., & HarrisonM, M. (2012). Adding sense to the Internet of Things: An architecture framework for smart object systems. *Personal and Ubiquitous Computing*, 16(3), 291-308. DOI: 10.1007/s00779-011-0399-8.
5. Shukla, V. K., & Singh, B. (2019, February). Conceptual framework of smart device for smart home management based on RFID and IoT. In *2019 Amity International Conference on Artificial Intelligence (AICAI)* (pp. 787-791). IEEE. DOI: 10.1109/AICAI.2019.8701301.
6. Srivastava, M., Abdelzaher, T., & Szymanski, B. (2012). Human-centric sensing. *Philosophical Transactions of the Royal Society A: Mathematical, Physical and Engineering Sciences*, 370(1958), 176-197.
7. Kulkarni, R. V., Förster, A., & Venayagamoorthy, G. K. (2010). Computational intelligence in wireless sensor networks: A survey. *IEEE Communications Surveys & Tutorials*, 13(1), 68-96. DOI: 10.1109/SURV.2011.040310.00002
8. Boberski, M. (2010). The ten most critical Web application security risks. *Techique Report, OWASP Foundation*.
9. Claycomb, W. R., & Nicoll, A. (2012, July). Insider threats to cloud computing: Directions for new research challenges. In *2012 IEEE 36th Annual Computer Software and Applications Conference* (pp. 387-394). IEEE.
10. Biswas, K., & Muthukkumarasamy, V. (2016, December). Securing smart cities using blockchain technology. In *2016 IEEE 18th international conference on high performance computing and communications; IEEE 14th international conference on smart city; IEEE 2nd international conference on data science and systems (HPCC/SmartCity/DSS)* (pp. 1392-1393). IEEE.
11. J. Burton (2019). The facts about Big data storage in smart cities applications. https://www.securityinfowatch.com/video-surveillance/video-surveillance- storage/article/21089531/seagate-the-facts-about-big-data-storage-in-smart-cities-applications
12. Christidis, K., & Devetsikiotis, M. (2016). Blockchains and smart contracts for the internet of things. *IEEE Access*, 4, 2292-2303.
13. Gupta, R., Shukla, V. K., Rao, S. S., Anwar, S., Sharma, P., & Bathla, R. (2020, January). Enhancing privacy through "smart contract" using blockchain-based dynamic access control. In *2020 International Conference on Computation, Automation and Knowledge Management (ICCAKM)* (pp. 338-343). IEEE. DOI: 10.1109/ICCAKM46823.2020.9051521.
14. Anwar, S., Shukla, V. K., Rao, S. S., Sharma, B. K., & Sharma, P. (2019, November). Framework for financial auditing process through blockchain technology, using identity based cryptography. In *2019 Sixth HCT Information Technology Trends (ITT)* (pp. 099-103). IEEE. DOI: 10.1109/ITT48889.2019.9075120.
15. Selinger, M. (2015). Test: Fitness wristbands reveal data. *AV-Test: The Independent IT- Security Institute*.

16. Treiblmaier, H., Rejeb, A., & Strebinger, A. (2020). Blockchain as a driver for smart city development: application fields and a comprehensive research agenda. *Smart Cities*, 3(3), 853-872.

17. Liao, D. Y., & Wang, X. (2018, December). Applications of blockchain technology to logistics management in integrated casinos and entertainment. In *Informatics* (Vol. 5, No. 4, p. 44). Multidisciplinary Digital Publishing Institute.

18. Nayyar, A. (2019). *Handbook of Cloud Computing: Basic to Advance research on the concepts and design of Cloud Computing*. BPB Publications.

19. Shaikh, R., & Sasikumar, M. (2012). Security issues in cloud computing: A survey. *International Journal of Computer Applications*, 44(19), 4-10.

20. Kumar, S., & Goudar, R. H. (2012). Cloud computing-research issues, challenges, architecture, platforms and applications: a survey. *International Journal of Future Computer and Communication*, 1(4), 356.

21. SaaS vs PaaS vs IaaS: What's The Difference & How To Choose. https://www.bmc.com/blogs/saas-vs-paas-vs-iaas-whats-the-difference-and-how-to-choose/#ref2

22. Rajesh, S., Swapna, S., & Reddy, P. S. (2012). Data as a service (daas) in cloud computing. *Global Journal of Computer Science and Technology*, 12(11), 25-29.

23. Zhang, S., Zhang, S., Chen, X., & Huo, X. (2010, January). Cloud computing research and development trend. In *2010 Second International Conference on Future Networks* (pp. 93-97). IEEE.

24. Voas, J., & Zhang, J. (March/April 2009). Cloud computing: New wine or just a new bottle? *IEEE ITPro*, 15–17.

25. Rathore, M. M., Paul, A., Ahmad, A., & Jeon, G. (2017). IoT-based big data: from smart city towards next generation super city planning. *International Journal on Semantic Web and Information Systems (IJSWIS)*, 13(1), 28-47.

26. Peng, S., Wang, G., & Xie, D. (2016). Social influence analysis in social networking big data: Opportunities and challenges. *IEEE Network*, 31(1), 11-17.

27. Jung, J. J. (2017). Computational collective intelligence with big data: Challenges and opportunities. *Future Generation Computer Systems*, 66, 87-88.

28. Narman, H. S., Hossain, M., Atiquzzaman, M., & Shen, H. (2017). Scheduling internet of things applications in cloud computing. *Annals of Telecommunications*, 72(1), 79-93. DOI: 10.1007/s12243-016-0527-6

29. Liu, B., Yu, X. L., Chen, S., Xu, X., & Zhu, L. (2017, June). Blockchain based data integrity service framework for IoT data. In *2017 IEEE International Conference on Web Services (ICWS)* (pp. 468-475). IEEE. DOI: 10.1109/ICWS.2017.54

30. Nepal, S., Chen, S., Yao, J., & Thilakanathan, D. (2011, July). DIaaS: Data integrity as a service in the cloud. In *2011 IEEE 4th International Conference on Cloud Computing* (pp. 308-315). IEEE. DOI: 10.1109/CLOUD.2011.35

31. Puar, V. H., Bhatt, C. M., Hoang, D. M., & Le, D. N. (2018). Communication in internet of things. In *Information Systems Design and Intelligent Applications* (pp. 272-281). Springer, Singapore. DOI: 10.1007/978-981-10-7512-4_28

32. Singh, D. K., Sobti, R., Jain, A., Malik, P. K., & Le, D. N. (2022). LoRa based intelligent soil and weather condition monitoring with internet of things for precision agriculture in smart cities. *IET Communications*, 16(5), 604-618.

33. Seth, B., Dalal, S., Le, D. N., Jaglan, V., Dahiya, N., Agrawal, A., ... & Verma, K. D. (2021). Secure Cloud Data Storage System Using Hybrid Paillier–Blowfish Algorithm. *CMC-Computers, Materials & Continua*, 67(1), 779-798.

34. Li, J., Li, Y. K., Chen, X., Lee, P. P., & Lou, W. (2014). A hybrid cloud approach for secure authorized deduplication. *IEEE Transactions on Parallel and Distributed Systems*, 26(5), 1206-1216. DOI: 10.1109/CESYS.2016.7889973.

35. Gueron, S., Johnson, S., & Walker, J. (2011, April). SHA-512/256. In *2011 Eighth International Conference on Information Technology: New Generations* (pp. 354-358). IEEE. DOI: 10.1109/ITNG.2011.69.

8

Role of IoT in Smart Homes and Offices

Shaurya Gupta[1], Sonali Vyas[1], Kanta Prasad Sharma[2]

[1] School of Computer Science, UPES Dehradun, Uttarakhand, India
[2] GLA University Mathura, India
Email: shaurya55@gmail.com, vyas.sonali86@gmail.com, tokpsharma@gmail.com

Abstract

The digitization of the physical world and obstacles in day-to-day life has opened up more avenues in terms of research and innovation for researchers for making our daily lives easier. The important technologies are either a part of cloud computing or the internet of things. Existing applications include scenarios in smart offices and homes. The key effort lies in the integration and control of the device services in a smart environment. The internet of things (IoT) is a technology that allows us to add a device to an inert object, such as physical microelectric hardware units and automobiles, at times covering entire structures over IP networks, enabling them to interconnect. All these devices are embedded with software, sensors and actuators for collecting and exchanging data amongst themselves. Cloud computing plays a crucial part in IoT as associated devices can attain supplementary services like servers, databases, software, networks, analytics and other computing functions that can be operated through the cloud. This chapter discusses the concept of smart offices and homes, which are a part of the smart building environment structure, and the role of IoT and cloud computing in establishing communication amongst devices. It further discusses the components of each technology and their areas of application, and future aspects and limitations.

Keywords: IoT, smart home, smart office, smart thermostat, smart sensor

8.1 Introduction

The digitization of the physical world and the obstacles in day-to-day life has opened up more avenues in terms of research and innovation for researchers to make our daily lives easier [1,2]. The important technologies are either a part of cloud computing or the internet of things (IoT). Existing applications include those for use in smart home and office scenarios. The key effort lies in the integration and control of the device services in a smart environment. IoT is defined as a group of hardware electronic devices, which have a particular IP address over an IP network that enables them to interconnect [3]. All these devices are embedded with software, sensors and actuators for collecting and exchanging data amongst themselves. Cloud computing is linked to the IoT as the connected devices need to attain supplementary resources like storage and computing power from cloud infrastructure services [4-6]. Various settings for homes, offices, and others associated with the IoT are presented in Table 8.1.

Table 8.1: Different IoT settings for smart homes, offices, etc.

S.No.	Settings	Purposes
1	Home	Chore automation and security
2	Offices	Security
3	Factories	Operations and equipment optimization
4	Retail Enviromnents	Automated checkout
5	Worksites	Operations optimization/health and safety
6	Human	Health and fitness
7	Outside	Logistics and navigation
8	Cities	Public health and transportation
9	Vehicles	Autonomous vehicles and condition-based maintenance

8.2 Smart Building Constituents

A smart building has the following constituents which interact amongst themselves to implement the concept of a smart city as a whole. Smart offices are environments which are intellectual, integrated and context-sensitive [7-8]. These environments consist of systems which interact with humans in adaptive, dynamic and inconspicuous ways to support routine office tasks [9, 10]. These settings are established by scrutinizing contextual data which is acquired by associated systems with the help of cameras or microphones integrated into the office environment [10]. Users interact with their environments using actuators which can be speakers, displays or automatic doors. The surroundings are quite supple as they possess the required skill to change cohesive systems which are completely based on composed analyzed contextual data [11]. Key dissimilarities amongst smart office and home surroundings depend on the environments in which they are functioning along with the set of devices in use for their specific purposes, and involves the concept of user's rights for access control in a very sophisticated manner.

Smart office surroundings are an exceptional sort of smart structure [12, 13] where there are many kinds of smart building environments which involve optimization of energy and resources in day-to-day lives. The organization of a smart building environment is shown in Figure 8.1.

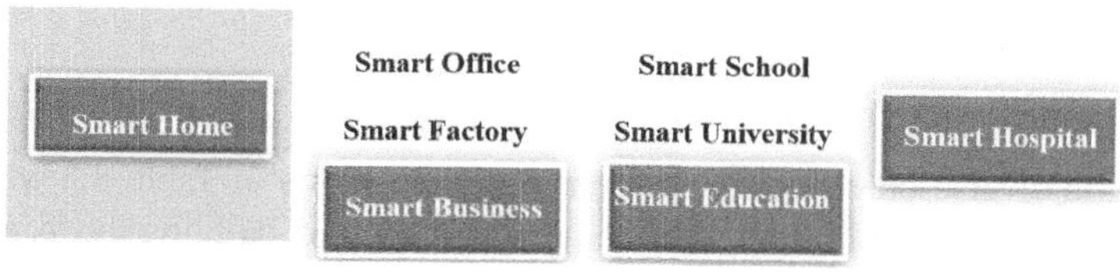

Figure 8.1: Constituents of a smart building.

Services [14-16] provided by smart building environments involve the types of services and objectives displayed in Table 8.2.

Table 8.2: Different amenities of a smart building.

Context	Services	Objectives
Smart Building	Water Management Parking System Fire/Smoke Detection and Alarm Energy and Information Management	Energy Efficiency Cost Optimization Safety Enhancements Security Enhancements

Environment-specific services delivered by varied smart structure surroundings along with their specific objectives are shown in Table 8.3.

Table 8.3: Services and goals of different smart building environments.

Smart Building Environments	Services	Objectives
Smart Home	Remote Control of Home Application Health Monitoring Multimedia System Control	Cost Optimization Energy Efficiency Comfort
Smart Office	Communication and Collaboration Systems Office Management System Personal Digital Assistance	Cost Optimization Energy Efficiency Process Optimization Time Management and Optimization Employee Satisfaction
Smart School and University	Communication and Collaboration Systems Information System and Services Asynchronous Teaching	Learning Process Optimization Motivational Learning for Students Ease of Learning Learning Anytime
Smart Hospital	Patient Experience Service Patient Registration System and Service	Cost Optimization Energy Efficiency Comfort Process Optimization Service Optimization Increase in Patient Satisfaction

The foremost objectives of a smart structure are:

- Energy and cost optimization

- Ease and comfort
- Enhancements in terms of security and safety

Hence, the diverse classes of smart constructions shown in Table 8.3 which improve and augment the aforesaid facets. More specifically, smart building environments aid in optimization of services and are paramount to client contentment overall. A completely functional smart home or office implies that IoT should be ubiquitous and acceptable in all junctions of a bounded area. In the case of homes and offices, a solo wireless access point is not at all satisfactory; therefore, a feasible resolution is to either ramp up communication control or to use devices which act as range extenders.

8.3 Concept of Smart Office Service Devices

Office hardware devices are incorporated into a mutual setting along with varied tasks that require automation to create a situation for functioning of services being presented by these devices. The automation of the devices in a smart office surrounding is depicted in Figure 8.2.

Figure 8.2: Shared service devices in smart office.

Taking an example of smart office, workers control the hardware units either via Amazon Alexa or a network calendar like Microsoft Outlook. A set of diverse instructions help in controlling devices of the smart office. Calendar appointments scheduled for any room, are responsible for triggering any action for the device (see Figure 8.3).

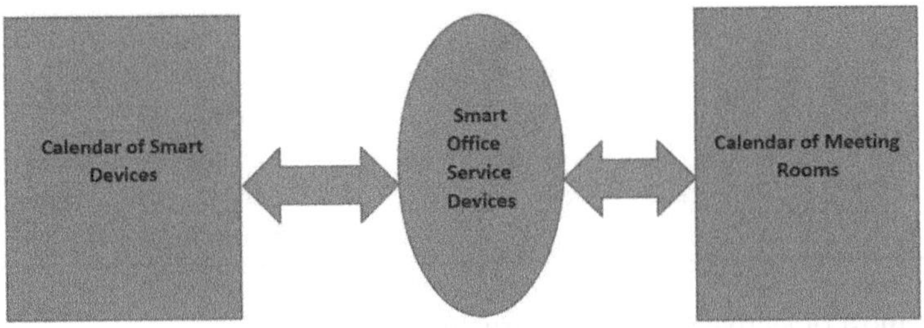

Figure 8.3: Smart office service device sharing.

The proper procedure should be used to synchronize the device with the calendar so that these devices can start up and shut down in an instant. Workers can define jobs in their smart devices or regulate them straightaway via generating calendar appointments or with the help of voice commands. Table 8.4 discusses some of the principal services and systems for smart offices.

Table 8.4: Principal services and systems for smart offices.

System	Short Description	Objectives
Parking Management	Concepts and systems to guide customers in locating a parking space in a confined area	Comfort gain Increase of customer satisfaction
Access Control	Intelligent access control, e.g., by face recognition.	Security enhancements
Device Management	Initial settings for devices, e.g., switch off lights when nobody is in the room. Personalized public devices locate devices closest to a certain user. e.g., a printer or scanner.	Comfort gain Energy efficiency Cost optimization
Communication Systems and Services	Software tools and hardware devices for online conferences, Services showing the availability of users. Supporting communication between distributed users.	Improvement of communication and collaboration
Collaboration Systems and Services	Concepts for intuitive and collaborative input and output interfaces services. Allows working together on documents with different kinds of devices.	Improvement of collaboration
Information Services	Information services can be reached at any time from anywhere by different flexible interaction concepts. Service for easy data access.	Comfort gain Time optimization Process optimization
Visitor Guidance	Flexible, easy and comfortable concepts to guide visitors within the office building, e.g., by automated doors.	Comfort gain Increase in visitor satisfaction
Room Management	Automated room allocation, e.g., locating an available meeting room closest to a certain user or recommendation of an available office that best fits the preferences of a certain user or group of users	Comfort gain Cost optimization Time optimization Process optimization

8.4 IoT in Smart Homes and Offices

The internet of things (IoT) is an assortment of smart devices and actuators along with cell phones capable of sharing system assets together. Device alliance and harmonization revolves around an actual network, and with the help of the internet is handled by servers managing cloud database. Software distinct designs [15, 16] are responsible for configuring ACs, TVs, smartphones, iPads, lighting facilities, and traffic monitoring systems which all constitute a device network or a network of devices. It even includes configuring a home security system which is managed by cloud database servers and virtual networks. Technologies involving software-defined architecture assist in energy management [17] for devices in a real-world application scenario. The multilayer design of IoT architecture provides a substantial right to the use of smart devices with a higher probability of device usability, improved connectivity, remote monitoring of devices, and also aids in overall network management of IoT hardware units.

The IoT structure inculcates the idea of context awareness in addition to standards for coordination of multiple users, distribution of network resources, and monitoring energy consumed by varied networks, as shown in Figure 8.4.

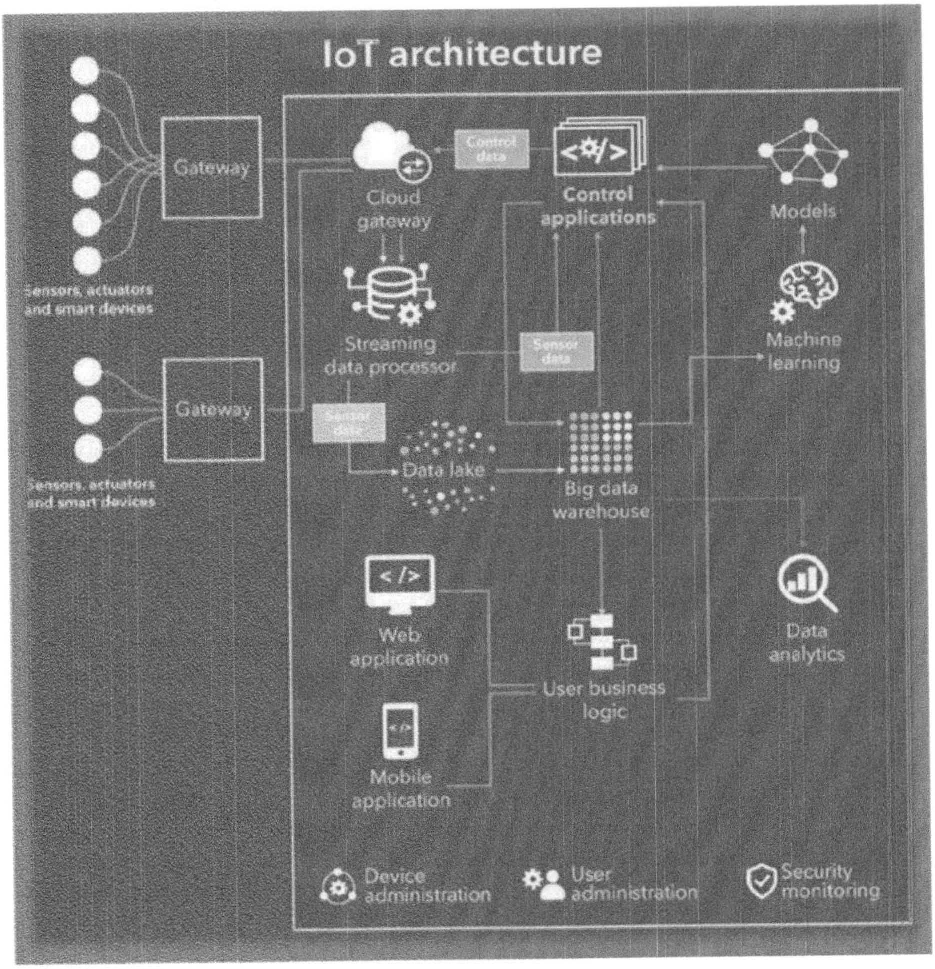

Figure 8.4: IoT architecture for smart homes and offices.

The operative characteristics of context awareness encompass the following:

- Classification of smart devices according to standard facility and enabling them to be user friendly.
- Smearing of the business rules means internal policy decision of any organization depending on the services of the user. Smart healthcare applications encompass IoT characteristics in sensors which are close to patients and therefore help in monitoring the patients' health to avoid any health threats.

Smart framework guarantees sophisticated functionality which is responsible for monitoring customer mobility configurations and preserves the integrity and privacy of user data. IoT incorporation involves unification of sensors with the benefit of image processing in the case of a framework suitable for smart homes [18]. Additionally, it allows synchronization of the devices of all family members in a home. The sensors incorporate resource sharing controlled by the centralized controller in a smart home environment. It empowers synchronization between the devices along with sharing of network properties. Centralized server is responsible for defining the serviceable characteristics of every device which is based on the following:

- Specifications
- Priority
- Energy resource.

In the case of a smart home [19], cognitive computing technology relies on practical methodology for real-time situations. It sustains faults in multiple utility procedures in the electrical power system for the smart home, which is depicted in Figure 8.5.

Figure 8.5: Smart home in conjunction with cloud server and third-party services.

Visual-based tracking systems, such as cameras, monitor the positions of inhabitants in a smart room. Artificial intelligence (AI)-centered IoT context [20] relates to vision-centered tracing structure, which identifies an individual on the basis of facial appearance in addition to sentiment analysis. Context awareness allows users to arrange services in addition to disclosing their location to cloud-native servers. Perceptual detection systems interpret the environment for the aging and vision-impaired with well-timed alarms in adjacent zones that alert them to any obstacles in their way when walking, and these signals are aired through monitoring devices. IoT incorporation [20] involves the process of integrating sensors having image processing ability. The device and network layer are embedded with AI procedures and the adaptive application layer protocol controls and also synchronizes adjacent surroundings.

8.4.1 Smart Environment Models

Smart environment models [21] effectively manage sensor nodes which are positioned at numerous locations and are responsible for analysis of sporadic data besides addressing concerns related to remote management of the network. A centralized controller is responsible for managing devices, including sensors, surveillance cameras and RFIDs. Smart home environment manages the overall energy use in the house, including the temperature and air pressure, and accurate data are delivered on the user's devices. The monitoring system in smart home [22] architecture involves monitoring the connectivity of devices at a physical layer. Visible light communication [23] powered by LED elements of sensor nodes deliver consistent amenities which are aimed at enclosed surroundings. Information management model [24] advances the interoperability of devices by means of categorizing user services besides describing the guidelines of data interchange through exterior database. A 3D-enabled GUI is used in the case of multifaceted IoT smart environments [24] and it contemplates the related perception rules for controlling IoT units. Smart lock systems [25-26] have the authority to control electronic control devices besides communicating with user smartphones or communication servers.

8.4.2 IoT Control Systems for Smart Home Devices

The IoT framework consists of associated units which enable amenities for inhabitants of a smart household. It follows the client-server model, wherein the client is an inhabitant of a smart household and with the help of a device like a smartphone can access the services and server, which acts as an integrated component which is responsible for processing requests of connected devices [27]. The IoT smart home structure has a client which panels associated devices with the help of a smartphone.

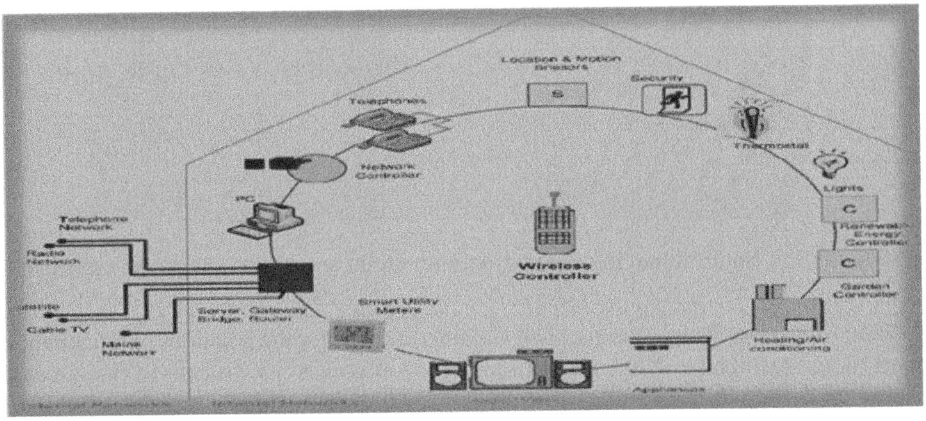

Figure 8.6: Outline of an IoT-enabled household.

Figure 8.6 specifies the framework of an IoT-based smart household which performs the following tasks:

- Smart thermostat sensors record the temperature and are responsible for sending a request to the principal AC entity, in addition to regulating the surrounding temperature. They are responsible for tracking weather conditions, then regulating temperature to optimum levels.
- Smart refrigerators, aided by cameras, classify foodstuff in addition to sending notifications to inhabitants of the smart household.
- Smart air conditioners, aided by motion and infrared sensors, are equipped with fans which help control their speed, which is completely based on the movements of the inhabitants in a confined area or room.

- Smart fans and lights switch on automatically whenever a person enters a room. Sensors logically control lighting in a smart household. Cameras recognize the presence of inhabitants, and consequently the smart devices are triggered.

- Smart TVs can send reminders to users about their favorite programs and can record them for later viewing.

- Smart water heaters are activated based on settings and choices specified by the user.

Smart IoT deployment into home components are well-ordered by a centralized server. Server middleware facilities form the links amongst IoT smart home components [28].

Figures 8.7 and 8.8 show the flow of data between server and connected smart home-based components. Event categorization is observed on the user's smartphone. The server is responsible for monitoring the state of devices, which can be either busy or idle, in addition to planning events based on the user profile and conditions in the smart home. Smart home-based devices optimize the use of resources in the smart home-based network.

A smartphone app helps the user identify the accessible devices present in a room. Identified controllable devices are displayed on the screen of a smartphone with their respective features. A standalone application controls numerous sorts of devices in such a way to dim lights or control the refrigerator thermostat. Scalability of systems is attained via the configuration of an XML file stored locally, which also has a list of all devices which can be discovered via commands or graphical user interface. The XML file provides the feature of scalability; therefore, users need not recompile the governing application to add a novel unit of the device [29].

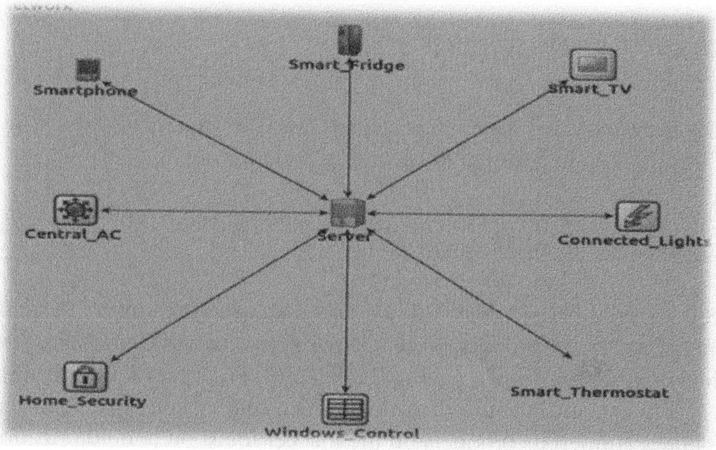

Figure 8.7: IoT smart household components deployment with central server.

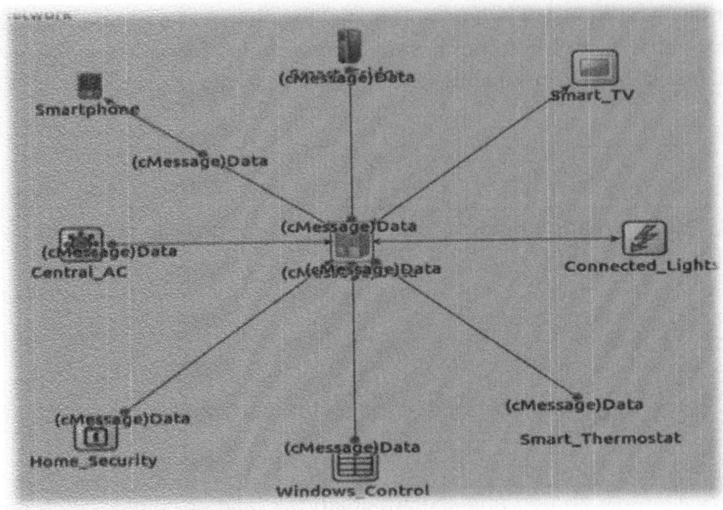

Figure 8.8: Data exchange amongst IoT smart home devices and server.

8.4.3 Prevailing Designs in Smart Homes

Universally, the current home and office computerization systems are categorized according to the following classes:

- Lighting control systems.
- Power on/off regulator switch for devices which are either remote or timer regulated.
- Temperature control at some distance with the help of a remote control which implements the concept of temperature regulating systems.
- Kitchen garden irrigation control systems.
- Home-based safety, security and reconnaissance arrangements which involve remote, native observation in addition to recording and alerting.
- Automation systems with custom design.

Figure 8.9 shows a classic conventional home automation control system [30]. These arrangements of device units collectively taken together have a limitation of purchasing every individual control unit from a solitary benefactor. For operative controlling of all units of devices in a household, the chief regulator core unit is managed by an app connected by a smartphone. Numerous products on the market focus on computerization of lighting, temperature control, management of entertainment-based devices and, lastly, the security and safety systems [31-34].

Figure 8.9: Custom home automation and control system.

8.4.4 Smart Home Constituents

- Server: A server is a vital monitoring and managing system backed up by a database. The server enables remote entry into connected smart devices. A smartphone directs the request to the server, which validates the source node in addition to forwarding the message to the target node. The server applies encryption procedures for ensuring system security, in addition to denying unlicensed admittance to IoT framework.

- Home Security System: This framework embraces the functionality for adding new users, deleting existing users as well as enabling authorizations to legitimate users for accessing varied services in a smart household. A home-based security system is equipped with UV in addition to a thermal camera which recognizes any stranger in an adjoining area, in addition to locking the door if any unauthorized person rings the doorbell, and alerting the home inhabitants so they can decide whether to open the door or not [35].

- Smart Thermostat: In a smart home-based network it controls the temperature and also regulates the crucial AC unit to ideal levels based on a particular room temperature to the temperature outside the house. It guarantees that if there is no single individual present in house, the associated smart hardware devices are switched to a power saving mode for the purpose of saving energy.

- Smart Air Conditioner: A smart air conditioner controls room temperature and also regulates air outflow in the house. If any inhabitant needs to have cooling in any room or specific area of the house, then the AC apertures start functioning for that specific area. It routes the cool air to the essential location in addition to maintaining optimal room temperature. Infrared sensors which are implanted throughout the smart home are responsible for detecting the presence of inhabitants in the house, which results in energy saving [35].

- Smart TV: A smart TV is an in-house unit in a smart home-based system which is linked to the central server and supports online video streaming. Inhabitants of the smart home use a smartphone for supervising the playback and rostering favorite program footage.

8.4.5 Cloud Topological Structure

This structure, which is based on the concept of a smart home, is a revision of the cloud arrangement with the addition of smart home infrastructure and integration of middleware services into the cloud management platform.

- Enterprise Scale Public Cloud: It is very vital for the entire cloud architectural design for ensuring that the cloud works efficiently without the support of a smart home.
- Third-Party Cloud: Services delivered by enterprise-scale public cloud are essential. Consequently, cloud architectural design permits a third party to deploy certain applications for specific users by implementing a platform-based architecture like Google App Engine [36].
- Smart Home Cloud: It is virtualized as a node at the home gateway layer. Various nodes arrange themselves into clusters, which in turn form the constituents of smart home-based cloud, which is a fragment of the entire cloud architecture that is quite comparable to other clouds shaped by clusters of computers. Home gateway is a momentous part of the smart home-based cloud-like computer operating system. It covers services and provides methods for permitting devices outside the smart home to utilize them, in addition to also being responsible for searching out resources and facilitating communication amongst home appliances (see Figure 8.10).

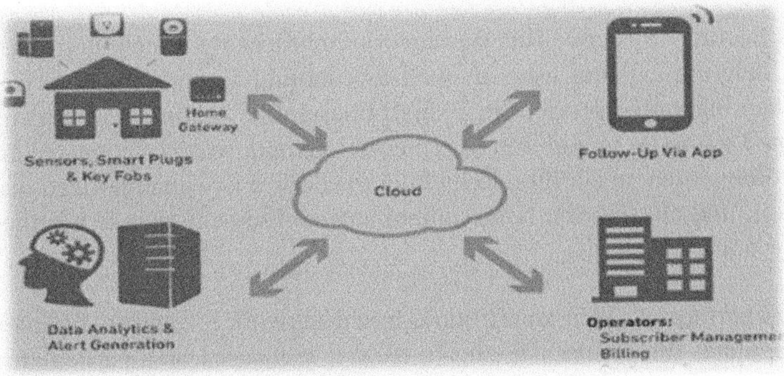

Figure 8.10: Cloud architecture for smart home.

8.4.6 Smart Home-Based Cloud Architectural Design

A smart home-based cloud architectural design is established by considering the three constituents of a smart home: infrastructure layer, platform layer, and service layer.

- Infrastructure Layer: It is the lowest level in cloud design architecture which consists of virtualized and physical assets. The virtualization component is responsible for virtualizing resources which cloud can regulate and it further includes:
  - Computational power
  - Storing or storage space
  - Network assets

- Platform Layer: It is responsible for managing resources and security module. The resource management module schedules resources in addition to detecting the status of system processing and also cataloging and eliminating components of a virtualized smart house. The other component manages and guarantees reliability of cloud dispensation plus guards the monetary privileges of third parties vending their services [36-45].
- Service Layer: It is responsible for providing services to third-party developers who are responsible for crafting, in addition to deploying, applications and services which are provided by application programming interfaces of the cloud platform.

8.5 Future Research Directions and Limitations of Smart Home-Based Technology

Remarkable studies are being carried out on ways of implementing smart home-based technologies, though they are still in the initial phase. Various complications and challenges still require attention. The key challenges are:

- Integration of power efficiency in IoT architecture for achieving remarkable performance in the case of smart home-based devices.
- Energy consumption models must be reliable with terms and conditions of smart homes and offices.
- Communicating devices should use protocols which are energy efficient with less power intake.
- Efficient cloud management with respect to power consumption.

Smart home-based technology will ultimately drive the ecosphere to sustainability, but it needs to comply with laws and regulations in order to achieve better productivity for the organization. Therefore, it is fairly valuable for society. The actual protagonists in this are the home/organization front-runners who trust in ideologies of smart technology and follow these moralities whilst making the businesses productive.

8.6 Conclusion

This chapter discussed the smart home and its functionality in an urban scenario, in addition to its constituents, the role of IoT in smart home and offices, the context of smart buildings and their services being delivered, and the roles of service devices in any smart office. Furthermore, the IoT architecture for smart homes and offices was discussed in conjunction with cloud servers and third-party services. Lastly, it deliberated on contemporary designs for smart homes, in addition to cloud topological structures existing in smart homes and offices.

References

1. Manyika, J., Chui, M., Bisson, P., Woetzel, J., Dobbs, R., Bughin, J., & Aharon, D. (2015). *The Internet of Things: Mapping the value beyond the hype* (Vol. 24). New York, NY, USA: McKinsey Global Institute.

2. Baller, S., Dutta, S., & Lanvin, B. (2016). *Global information technology report 2016*. Geneva: Ouranos.
3. Rose, K., Eldridge, S., & Chapin, L. (2015). The internet of things: An overview. *The Internet Society (ISOC)*, 80, 1-50.
4. Serrano, M., Quoc, H. N. M., Hauswirth, M., Wang, W., Barnaghi, P., & Cousin, P. (2013, June). Open services for IoT cloud applications in the future internet. In *2013 IEEE 14th International Symposium on A World of Wireless, Mobile and Multimedia Networks (WoWMoM)* (pp. 1-6). IEEE.
5. Razvi, S. A. M., Al-Dhelaan, A., Al-Rodhaan, M., & Sulaiman, R. A. B. (2015). IoT cloud-sensor secure architecture for smart home. In *Proceedings of the International Conference on Security and Management (SAM)* (p. 243). The Steering Committee of The World Congress in Computer Science, Computer Engineering and Applied Computing (WorldComp).
6. Sivakumar, S., Anuratha, V., & Gunasekaran, S. (2017). Survey on integration of cloud computing and internet of things using application perspective. *International Journal of Emerging Research in Management & Technology*, 6(4), 101-108.
7. Röcker, C. (2010, January). Services and applications for smart office environments-a survey of state-of-the-art usage scenarios. In *Proceedings of the International Conference on Computer and Information Technology (ICCIT 2010)*, Cape Town, South Africa (pp. 1173-1189).
8. Gal, C. L., Martin, J., & Durand, G. (2000). Smart office: an intelligent and interactive environment. In *Managing Interactions in Smart Environments* (pp. 104-113). Springer, London.
9. Ramos, C., Marreiros, G., Santos, R., & Freitas, C. F. (2010). Smart offices and intelligent decision rooms. In *Handbook of Ambient Intelligence and Smart Environments* (pp. 851-880). Springer, Boston, MA.
10. Nakashima, H., Aghajan, H., & Augusto, J. C. (Eds.). (2009). *Handbook of ambient intelligence and smart environments*. Springer Science & Business Media.
11. Mikulecký, Peter. Smart environments for smart learning. *DIVAI 2012 9th International Scientific Conference on Distance Learning in Applied Informatics*. 2012.
12. Bhuyar, R., & Ansari, S. (2016). Design and implementation of smart office automation system. *International Journal of Computer Applications*, 151(3), 37-42.
13. Rottondi, C., Duchon, M., Koss, D., Verticale, G., & Schätz, B. (2015, March). An energy management system for a smart office environment. In *2015 international conference and workshops on networked systems (NetSys)* (pp. 1-6). IEEE.
14. Horch, A., Kubach, M., Roßnagel, H., & Laufs, U. (2017, November). Why should only your home be smart?-a vision for the office of tomorrow. In *2017 IEEE International Conference on Smart Cloud (SmartCloud)* (pp. 52-59). IEEE.
15. Granjal, J., Monteiro, E., & Silva, J. S. (2015). Security for the internet of things: a survey of existing protocols and open research issues. *IEEE Communications Surveys & Tutorials*, 17(3), 1294-1312.
16. Cicirelli, F., Fortino, G., Giordano, A., Guerrieri, A., Spezzano, G., & Vinci, A. (2016). On the design of smart homes: A framework for activity recognition in home environment. *Journal of Medical Systems*, 40(9), 1-17.
17. Kirkham, T., Armstrong, D., Djemame, K., & Jiang, M. (2014). Risk driven Smart Home resource management using cloud services. *Future Generation Computer Systems*, 38, 13-22.
18. Belley, C., Gaboury, S., Bouchard, B., & Bouzouane, A. (2015). Nonintrusive system for assistance and guidance in smart homes based on electrical devices identification. *Expert Systems with Applications*, 42(19), 6552-6577.
19. Mano, L. Y., Faiçal, B. S., Nakamura, L. H., Gomes, P. H., Libralon, G. L., Meneguete, R. I., ... & Ueyama, J. (2016). Exploiting IoT technologies for enhancing Health Smart Homes through patient identification and emotion recognition. *Computer Communications*, 89, 178-190.

20. Moser, K., Harder, J., & Koo, S. G. (2014, October). Internet of things in home automation and energy efficient smart home technologies. In *2014 IEEE International Conference on Systems, Man, and Cybernetics (SMC)* (pp. 1260-1265). IEEE.

21. Chan, M., Estève, D., Escriba, C., & Campo, E. (2008). A review of smart homes—Present state and future challenges. *Computer Methods and Programs in Biomedicine*, 91(1), 55-81.

22. Kaldeli, E., Warriach, E. U., Lazovik, A., & Aiello, M. (2013). Coordinating the web of services for a smart home. *ACM Transactions on the Web (TWEB)*, 7(2), 1-40.

23. Tiwari, S. V., Sewaiwar, A., & Chung, Y. H. (2017). Smart home multi-device bidirectional visible light communication. *Photonic Network Communications*, 33(1), 52-59.

24. Capitanelli, A., Papetti, A., Peruzzini, M., & Germani, M. (2014). A smart home information management model for device interoperability simulation. *Procedia CIRP*, 21, 64-69.

25. Nazari Shirehjini, A. A., & Semsar, A. (2017). Human interaction with IoT-based smart environments. *Multimedia Tools and Applications*, 76(11), 13343-13365.

26. Ho, G., Leung, D., Mishra, P., Hosseini, A., Song, D., & Wagner, D. (2016, May). Smart locks: Lessons for securing commodity internet of things devices. In *Proceedings of the 11th ACM on Asia Conference on Computer and CommunicationsSecurity* (pp. 461-472).

27. Granjal, J., Monteiro, E., & Silva, J. S. (2015). Security for the internet of things: a survey of existing protocols and open research issues. *IEEE Communications Surveys & Tutorials*, 17(3), 1294-1312.

28. Fortino, G., Giordano, A., Guerrieri, A., Spezzano, G., & Vinci, A. (2015, December). A data analytics schema for activity recognition in smart home environments. In *International Conference on Ubiquitous Computing and Ambient Intelligence* (pp. 91-102). Springer, Cham.

29. Kirkham, T., Armstrong, D., Djemame, K., & Jiang, M. (2014). Risk driven Smart Home resource management using cloud services. *Future Generation Computer Systems*, 38, 13-22.

30. Alves, F. A. M., & Thangaraj, C. (2016, November). A scalable modular heterogeneous system for home and office automation. In *2016 IEEE MIT Undergraduate Research Technology Conference (URTC)* (pp. 1-4). IEEE.

31. McRoberts, L. M., Paczkowski, L. W., & Rondeau, D. E. (2016). U.S. Patent No. 9,282,898. Washington, *DC: U.S. Patent and Trademark Office.*

32. Bhatia, S., Bajaj, J., & Roja, M. M. (2014). Technology, Systems and Implementation of a Smart Home Automation System: A Review. *International Journal of Computer Technology and Applications*, 5(5), 1690-1695.

33. Balog, M., Harris, M., & Buchert, R. (2016). U.S. Patent No. 9,246,757. Washington, DC: U.S. Patent and Trademark Office. *Commissioning devices for automation systems.*

34. Dobyns, K. P., Waldo, G. J., & Kreitzer, R. R. (2012). *RF input/output connector receptacle and control buttons for a measurement instrument.* U.S. Patent Application No. 29/391,244.

35. Krishna, M. B., & Verma, A. (2016, December). A framework of smart homes connected devices using Internet of Things. In *2016 2nd International Conference on Contemporary Computing and Informatics (IC3I)* (pp. 810-815). IEEE.

36. Yang, Y., Wei, Z., Jia, D., Cong, Y., & Shan, R. (2010, March). A cloud architecture based on smart home. In *2010 Second International Workshop on Education Technology and Computer Science* (Vol. 2, pp. 440-443). IEEE.

37. Srivastava, S., Saxena, S., Buyya, R., Kumar, M., Shankar, A., & Bhushan, B. (2021). CGP: cluster-based gossip protocol for dynamic resource environment in cloud. *Simulation Modelling Practice and Theory*, 108, 102275.

38. Chithaluru, P., Al-Turjman, F., Kumar, M., & Stephan, T. (2020). I-AREOR: An energy-balanced clustering protocol for implementing green IoT in smart cities. *Sustainable Cities and Society*, 61, 102254.

39. Bhushan, S., Kumar, M., Kumar, P., Stephan, T., Shankar, A., & Liu, P. (2021). FAJIT: a fuzzy-based data aggregation technique for energy efficiency in wireless sensor network. *Complex & Intelligent Systems*, 7(2), 997-1007.

40. Aggarwal, A., & Kumar, M. (2021). Image surface texture analysis and classification using deep learning. *Multimedia Tools and Applications*, 80(1), 1289-1309.

41. Punia, S. K., Kumar, M., Stephan, T., Deverajan, G. G., & Patan, R. (2021). Performance analysis of machine learning algorithms for big data classification: Ml and ai-based algorithms for big data analysis. *International Journal of E-Health and Medical Communications (IJEHMC)*, 12(4), 60-75.

42. Bhardwaj A., Al-Turjman F., Kumar M, Stephan T., Mostarda L., "Capturing-the- Invisible (CTI): A Novel Approach for Behavior-based Attacks Recognition in Industrial Control Systems" *IEEE Access*, Vol.8, No.1, 2020, pp. 104956- 104966.

43. Le, D. N., Bhatt, C., & Madhukar, M. (Eds.). (2019). *Security designs for the cloud, IoT, and social networking*. John Wiley & Sons.

44. Doss, S., Paranthaman, J., Gopalakrishnan, S., Duraisamy, A., Pal, S., Duraisamy, B., & Le, D. N. (2021). Memetic optimization with cryptographic encryption for secure medical data transmission in IoT-based distributed systems. *Computers, Materials & Continua*, 66(2), 1577-1594.

45. Pal, S., Díaz, V. G., & Le, D. N. (Eds.). (2020). *IoT: Security and privacy paradigm*. CRC Press.

9

Role of IoT in the Prevention of COVID-19

Ankit Saxena[1], Akash Sanghi[1], Swapnesh Taterh[2], Neeraj Bhargava[3]

[1] Department of CSE, Invertis University, Bareilly, India
[2] AIIT, AMITY University, Jaipur, India
[3] Department of CSE, MDS University, Ajmer
Email: ankit.saxena5@yahoo.com, swapnesh@hotmail.com, profneerajbhargava@gmail.com

Abstract

Today the challenge before the world is how to stop the expansion of the COVID-19 virus, as the world is currently facing the third wave of COVID-19, against which the World Health Organization (WHO) is regularly updating all of us to take precautions. To prevent the spread of the virus, individuals testing positive for the disease should be placed in isolation. Many countries are currently affected and the whole world is taking precautions by using the various methods as per the WHO guidelines. Crowded places, close-contact settings, and confined, enclosed spaces with poor ventilation, and touching the same place that multiple people have touched are the main causes of COVID-19 spreading. Nowadays, all the electronics components are moving toward atomization, which is completely dependent upon the IoT. In this chapter, we will focus on the "touch" cause of expansion with a deep study using the touchless technology concept of IoT sensor to address this cause of COVID-19 spreading, for which various simple networking components are detailed which can be used to perform many day-to-day electronics tasks remotely. The IoT technology is an up-and-coming technology that can be more beneficial in this pandemic environment. With the help of IoT networking components any device can be operated remotely.

Keywords: IoT, touchless switch, barrier for COVID-19, multipurpose COVID-19, COVID-19 virus-free switch

9.1 Introduction

Today the most challenging problem in the world is how to protect each person against COVID-19. On January 30, 2020, WHO declared COVID-19 a public health emergency of international concern. To stop the spread of the virus, individuals testing positive for the disease were placed in isolation. Discharge required clinical recovery with two negative sequential RT-PCR results within 24 hours, which was later updated to 10 days, after symptom onset, plus a minimum of 3 days without symptoms for symptomatic patients and 10 days after a positive test for asymptomatic patients [1]. The COVID-19 pandemic made us clearly aware of the major world challenges that we collectively face today [2]. Now the crisis – whether economic, political, environmental or social, etc. – tends to hit the poorest and weakest the hardest, laying bare the most acute societal and political weaknesses of countries around the world.

Till date there is no vaccine available to permanently cure or permanently prevent COVID-19 [3]. Most of the COVID-19 patients have mild to moderate symptoms, but some of them progress to severe pneumonia and some eventually develop acute respiratory distress syndrome, septic shock or multiple organ failure [4].

Discharge required clinical recovery with two negative sequential RT-PCR results within 24 hours, which was later updated to 10 days, after symptom onset plus a minimum of 3 days without symptoms for symptomatic patients and 10 days after a positive test for asymptomatic patients [1]. Research analysis concluded that the patients with severe disease frequently had an increased IgG response, which was associated with a worse outcome. The cause of antibody-dependent enhancement of SARS-CoV-2 infection is a major concern. A potential pathogenic effect of antibodies targeted at severe acute respiratory syndrome coronavirus 2 would be a major issue for vaccine development and antibody-based therapies. Additional independent large cohort studies are needed to substantiate or dismiss this possibility [4].

The clinical treatment is basically based on symptomatic management and oxygen therapy, with mechanical ventilation for COVID-19 effected patients who are in respiratory failure. Many antiviral drugs, such as nucleotide analogue remdesivir, are not approved for COVID-19 treatment. Now we need a vaccine that will directly affect the virus to cure the patient.

There is only one solution available, which is to follow the WHO guidelines for prevention. On analyzing these guidelines, we find that the main basic cause of COVID-19 spreading is touching any surface randomly by various persons. As a solution to restrict the spread, a touchless electric multipurpose board was made, which will successfully provide the electricity supply for any electronic equipment without touching it. With this board, anyone can operate the equipment from a safe distance without touching it, which works as a barrier to the spread of COVID-19.

9.2 A Modern Era Problem

COVID-19, also known as coronavirus, is a newly discovered disease [3]. This new virus, which was unknown before the outbreak began in Wuhan, China, in December 2019, continues to have a devastating effect on many countries.

Currently, there are some proper medicines or vaccines available for its treatment. The elderly and others having any medical problems, such as cardiovascular disease, chronic respiratory problems, diabetes and cancer, are more likely to develop serious illness. Since the COVID-19 virus spreads mainly through droplets of saliva or discharge from the nose when any virus-infected person coughs or sneezes, it's important to practice respiratory etiquette. The graph below shows the number of COVID-19 cases in India confirmed daily. The data is provided by a website analysis report [5].

Figure 9.1 shows that the growth in the number of patients is very high.

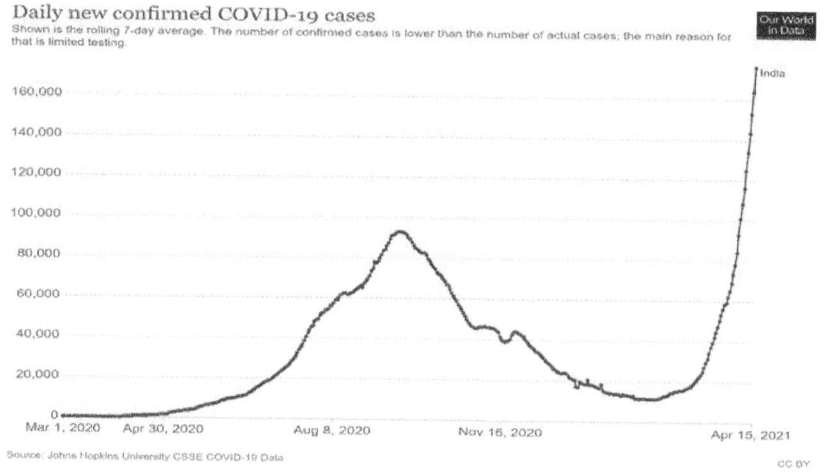

Figure 9.1: Growth in the number of COVID-19 patients in India.

Causes of COVID-19 [3]:

1. Contracting COVID-19 from others who are already infected.
2. It primarily spreads through small droplets from the mouth or nose , which are expelled when a patient with COVID-19 coughs, sneezes, or speaks.
3. It only is spread by someone who is infected and cannot develop automatically in any human.
4. The virus that causes COVID-19 can land on surfaces; therefore, it's possible to become infected by touching those surfaces and then touching the nose, mouth, or eyes.

Precautions against COVID-19 [3]:

1. Regularly wash your hands or sanitize them.
2. Regularly sanitize a surface which is regularly used by a lot of people.
3. Regularly sanitize the equipment which is generally used by many people.
4. Maintain social distancing (at least 1 meter distance) between yourself and others.
5. Avoid touching eyes, nose and mouth.

9.2.1 Current Risk

On January 30, 2020, WHO declared COVID-19 to be a public health emergency of international concern. To prevent the spread of the virus, it was determined that everyone individually testing positive for the disease should be placed in isolation. Discharge required clinical recovery with two negative sequential RT-PCR results within 24 hours, which was later updated to 10 days after symptom onset plus a minimum of 3 days without symptoms for symptomatic patients and 10 days after a positive test for asymptomatic patients [1]. Till date no vaccine is available to permanently cure or permanently prevent COVID-19. Very few countries have sufficient and appropriate diagnostic capabilities and obvious challenges exist to solve such outbreaks. Recent studies indicate that patients greater than 60 years of age are at higher risk than children who might be less likely to become infected or, if so, may show milder symptoms or a asymptomatic [6,7].

Table 9.1 was created by Eka Kotebe Treatment Center in Ethiopia; the risk analysis took into account several factors such as age, sex, etc. [8].

Table 9.1: Crude statistics relating to demographic risk factors for COVID-19 infection at Eka Kotebe Treatment Center in Ethiopia, October, 2020

Factors	Risk		COR (95% CI)	p value
	High (%)	Low (%)		
Age				
<25	34(77.3)	10(22.7)	3.40(0.592–19.541)	0.170
25–34	192(79)	51(21)	3.63(0.711–18.515)	0.121
35–44	14(56)	11(44)	1.27(0.214–7.581)	0.791
>45	3(50)	3(50)	1.00	
Sex				
Male	134(76.1)	42(23.9)	1.035(0.615–1.744)	0.896
Female	109(76.8)	33(23.2)	1.00	
Years of experience				
<2 years	67(74.4)	23(25.6)	1.515(0.667–3.442)	0.321
2–5 years	133(78.7)	36(21.3)	1.933(0.906–4.125)	0.088
>5 years	25(65.8)	13(34.2)	1.00	
Role of healthcare workers				
Medical doctor	44(86.3)	7(13.7)	4.746(1.921–11.728)	0.001***
Nurse	132 (84.1)	25 (15.9)	3.987(2.179–7.295)	0.000 ***
Health officer	11(84.6)	2 (15.4)	4.153(0.868–19.881)	0.075
Pharmacist	7(63.6)	4 (36.4)	1.321(0.360–4.851)	0.674
Other'	49(57.0)	37(43.0)	1.00	
Working department				
Inpatient	144(81.8)	32 (18.2)	3.044(1.584–5.850)	0.001***
Intensive care unit	52 (85.2)	9(14.8)	3.908(1.615–9.457)	0.002***
Laboratory	5(83.3)	1 (16.7)	3.382(0.371–30.872)	0.280
Pharmacy	7 (41.2)	10 (58.8)	0.474(0.157–1.424)	0.183
Other†	34(59.6)	23 (40.4)	1.00	
Work hours per day				
≤8 hours	220 (94.8)	12 (5.2)	1.00	0.007***
>8 hours	58 (84.1)	11(15.9)	3.309(1.395–7.884)	
Educational level				
Primary/Secondary	18(43.9)	23(56.1)	0.326(0.097–1.096)	0.070
Certificate/Diploma	14 (63.6)	8(36.4)	0.729(0.188–2.834)	0.648
MD/BSC	199 (83.6)	39(16.4)	2.126(0.709–6.376)	0.178
Specialty/ MSC/MPH	12 (70.6)	5(29.4)	1.00	
Underlying disease or pre-existing condition(s)				
Yes	10(62.5)	6(37.5)	0.494(0.173–1.406)	0.186
No	233(77.2)	69(22.8)	1.00	

Notes: Significant at: ***$p \leq 0.01$, 1=Reference category. 'Psychiatrist, nutritionist, midwife, laboratory personnel, x-technician, biomedical engineer, environmental health officer, anesthetist, social worker, clinical psychologist, porter, data manager, cleaner and spray man. †Psychiatric department, operation room, radiology, management team and IPC department.

Table 9.1 shows the risk analysis or the risk level for every worker who was performing duties in this pandemic situation. The main interaction was between health workers and COVID-19 patients, which put them at risk. Other departments were also important in the treatment of COVID-19 as they directly or indirectly interacted with COVID-19 patients.

The table shows the data based on the study analysis of the Eka Kotebe Treatment Center in Ethiopia in October 2020 according to their working hours, etc. [8].

Table 9.2: Multivariate logistic regression analysis of the relative effect of demographic, IPC and types of exposure factors of COVID-19 infection at Eka Kotebe Treatment Center in Ethiopia, October 2020.

Variables	Risk		COR (95% CI)	AOR (95% CI)
	High	Low		
Role of HCWs				
Medical doctor	44	7	4.746(1.921–11.728)***	3.174(0.928–10.856)
Nurse	132	25	3.987(2.179–7.295)***	1.233(0.503–3.021)
Health officer	11	2	4.153(0.868–19.881)	1.452(0.212–9.938)
Pharmacist	7	4	1.321(0.360–4.851)	2.139(0.141–32.381)
Other	49	37	1.00	1.00
Working department				
Inpatient	144	32	3.044(1.584–5.850) ***	2.772(0.962–7.988)
Intensive care unit	52	9	3.908(1.615–9.457) ***	6.545(1.787–23.966) **
Laboratory	5	1	3.382(0.371–30.872)	1.069(0.054–21.204)
Pharmacy	7	10	0.474(0.157–1.424)	0.288(0.027–3.032)
Other	34	23	1.00	
Work hours per day				
≤8 hours	220	12	1.00	1.00
>8 hours	58	11	3.309(1.395–7.884) ***	9.224(1.997–42.613)***
Availability of PPE				
Yes	31	25	0.295(0.160–0.544) ***	0.318(0.135–0.748) ***
No	210	50	1.00	1.00
Availability of alcohol-based hand rub				
Yes	208	72	0.199(0.046–0.856) **	0.093(0.009–0.979) **
No	29	2	1.00	1.00
One-meter distance				
Yes	86	1	26.158(6.27–109.111) ***	20.633(3.879–109.755) **
No	153	74	1.00	1.00
Perform AGPs				
Yes	69	1	23.2640(3.163–171.100) ***	5.133(0.584–45.125)
No	171	74	1.00	1.00
Environment: frequently-touched surfaces				
Yes	152	24	35.865(4.896–262.732) ***	9.600(1.053–87.494)**
No	90	51	1.00	1.00
Prolonged face-to-face exposure				
Yes	152	24	3.589(2.069–6.225) ***	1.561(0.749–3.351)
No	90	51	1.00	1.00

Note: Significant at: **p ≤0.01, ***p ≤0.001, 1=Reference category.

9.2.2 Awareness by Social Media

In the current era, we know that the easiest and fastest way to broadcast any information to the whole world is the social media platform. At the start of the COVID-19 pandemic, people were using social media more than usual because they rely on news from online sources to search for health information for themselves and their loved ones.

The use of media platforms (Whatsapp, Facebook, Linkedin, Instagram, etc.) has come as a welcome relief during the ongoing COVID-19 global pandemic. It is thought that analyzing social media usage in the context of global health catastrophes like the COVID-19 pandemic should help disclose the global mental health toll. The US Census Bureau surveyed more than 42% of people and identified symptoms of depression and higher anxiety levels in December 2020, which was 11% higher than the previous year. The survey findings reported by the Hazarika Commission on illegal migrants from Bangladesh to Assam during the pandemic had results similar to those of COVID-19 mental stress worldwide. When the COVID-19 global health crisis struck, a telephone service supported by the Assam Police of India studied 239 callers in April 2020 and found that 46% had anxiety, 22% indicated depression symptoms, and 5% had suicidal thoughts. This was enough evidence to convince the Government of Assam to launch a countrywide remote mental health telephonic service to tackle mental health well-being. Physical activities could act as medicine for noncommunicable diseases. After easing lockdowns and restrictions on social distancing in December 2020, the telephonic service collected the data of 43,000 people and found that 9% of them had anxiety symptoms, 4% had depression, and more than 12% reported stress related to the health crisis posed by the COVID-19 pandemic (see Table 9.3).

Table 9.3: Cases and mortality rate of countries most affected by COVID-19 as of February 23, 2022 (second wave).

Country	Confirmed	Deaths	Case-Fatality	Deaths/100k Pop.
United States	28,188,571	500,244	1.8%	152.90
India	11,016,434	156,463	1.4%	11.57
Brazil	10,195,160	247,143	2.4%	117.99
United Kingdom	4,138,233	120,988	2.9%	181.97
Russia	4,130,447	82,255	2.0%	56.93
France	3,669,354	84,764	2.3%	126.54
Spain	3,153,971	67,636	2.1%	144.76
Italy	2,818,863	95,992	3.4%	158.84
Turkey	2,646,526	28,138	1.1%	34.18
Germany	2,399,499	68,363	2.8%	82.44
Colombia	2,229,663	58,974	2.6%	118.78
Argentina	2,069,751	51,359	2.5%	115.43
Mexico	2,043,632	180,536	8.8%	143.07
Poland	1,642,658	42,188	2.6%	111.08
Iran	1,582,275	59,572	3.8%	72.83
South Africa	1,504,588	49,150	3.3%	85.06
Ukraine	1,354,545	26,531	2.0%	59.46
Indonesia	1,288,833	34,691	2.7%	12.96
Peru	1,283,309	45,097	3.5%	140.98
Czechia	1,157,180	19,330	1.7%	181.92
Netherlands	1,075,425	15,372	1.4%	89.21
Pakistan	573,384	12,658	2.2%	5.96

The social media platform provides the content of social support to the public seeking health information. For those suffering from health anxieties and medical conditions, social media offers them the significant benefit of correct online information. Social support explains the perception and practice of how social networks care for and value those within the networks. It explains how social networks embed individuals into social obligations and communication networks. The most popular one is how the social network is supportive, and localization of health through sports activities is also helpful [7].

9.2.3 Review of Current Solutions

As per the guidelines of WHO, each government is taking care of the citizens of their respective nations. Some of these guidelines are [2,6]:

- Install thermal scanners at the entry of government buildings as feasible. Mandatory placing of hand sanitizers at the entry of government buildings. Those found having flu-like symptoms may be advised to take proper treatment/quarantine, etc.
- Meetings, if feasible, should be done through video conferencing.
- Avoid non-essential official travel.
- Undertake essential correspondence on official email and avoid sending files and documents to other offices, to the extent possible.
- Ensure proper cleaning and frequent sanitization of the workplace, particularly of the frequently touched surfaces.
- Ensure regular supply of hand sanitizers, soap and running water in the washrooms.

9.2.4 Current Treatment

Coronavirus disease 2019 (COVID-19), a newly emerged disease caused by SARS-CoV-2, has recently become pandemic. Most COVID-19 patients exhibit mild to moderate symptoms, but some progress to severe pneumonia and some develop acute respiratory distress syndrome, septic shock and/or multiple organ failure. Discharge requires clinical recovery with two negative sequential RT-PCR results within 24 hours, which was later updated to 10 days after symptom onset plus a minimum of 3 days without symptoms for symptomatic patients and 10 days after a positive test for asymptomatic patients [1]. The clinical treatment is basically based on symptomatic management and oxygen therapy, with mechanical ventilation for COVID-19 effected patients who are in respiratory failure. Many antiviral drugs, such as the nucleotide analogue remdesivir, are not approved for COVID-19 treatment. Now we need a vaccine that will directly affect the virus to cure the patient [4].

To prevent the spread of the virus, individuals testing positive for the disease should be placed in isolation, Discharge requires clinical recovery with two negative sequential RT-PCR results within 24 hours, which was later updated to 10 days after symptom onset plus a minimum of 3 days without symptoms for symptomatic patients and 10 days after a positive test for asymptomatic patients. In addition to isolation, quarantine ("separation of persons who are not ill, but who may have been exposed to an infectious agent or disease"), measures were introduced. Individuals identified as contacts (e.g., providing direct care without the use of personal protective equipment, having face-to face-contact within 1 m $>$15 minutes) of laboratory-confirmed cases require 14 days of quarantine from the last time they were exposed to the patient [1].

Table 9.4: Demographics and clinical data for participants and COVID-19 status: isolated and quarantined.

		Total (n=502) n (%)	Isolated (n=301) n (%)	Quarantined (n=201) n (%)	P
Age, mean (SD)		40.41 (14.47)	41.10 (14.64)	39.36 (14.18)	0.187
Duration of segregation (days), mean (SD)		18.06 (14)	22.25 (16.4)	11.79 (4.67)	<0.001
Sex	Females	252 (50.2%)	151 (50.2%)	101 (50.2%)	0.986
	Males	250 (49.8%)	150 (49.8%)	100 (49.8%)	
Marital Status	Single	93 (18.5%)	54 (17.9%)	39 (19.4%)	0.975
	Married	394 (78.5%)	238 (79.1%)	156 (77.6%)	
	Divorced	7 (1.4%)	4 (1.3%)	3 (1.5%)	
	Widowed	8 (1.6%)	5 (1.7%)	3 (1.5%)	
Education	Elementary or below	22 (4.4%)	15 (4%)	7 (3.5%)	0.204
	Middle	69 (13.7%)	42 (14%)	27 (13.4%)	
	High	195 (38.8%)	126 (41.9%)	69 (34.3%)	
	Undergraduate	159 (31.7%)	90 (29.9%)	69 (34.3%)	
	Postgraduate	57 (11.4%)	28 (9.3%)	29 (14.4%)	
Occupation	Retired	93 (18.5%)	58 (19.3%)	35 (17.4%)	0.289
	Employed (public)	116 (23.1%)	60 (19.9%)	56 (27.9%)	
	Employed (private)	104 (20.7%)	62 (20.6%)	42 (20.9%)	
	Self-employed	14 (2.8%)	11 (3.7%)	3 (1.5%)	
	Unemployed	169 (33.7%)	106 (35.2%)	63 (31.3%)	
	Student	6 (1.2%)	4 (1.3%)	2 (1%)	
Smoking status	Smoker	98 (19.5%)	55 (18.3%)	43 (21.4%)	0.135
	Nonsmoker	365 (72.7%)	217 (72.1%)	148 (73.6%)	
	Ex-smoker	39 (7.8%)	29 (9.6%)	10 (5%)	
Comorbidities	Yes	224 (44.6%)	134 (44.5%)	90 (44.8%)	0.955
	No	278 (55.4%)	167 (55.5%)	111 (55.2%)	
History of mental conditions	Yes	14 (2.8%)	9 (3%)	5 (2.5%)	0.738
	No	488 (97.2%)	292 (97%)	196 (97.5%)	

9.3 Technology

In this section, all the WHO guidelines were considered when working on the "touchless" technology. According to the COVID-19 prevention guidelines, we should avoid touching frequently touched surfaces without sanitizing our hands afterwards.

9.3.1 Touchless User Interface

This is the process of commanding the computer via body motion and gestures without physically touching a keyboard, mouse or screen.

9.3.2 IR Sensor Working Principle

There are different types of infrared transmitters depending on their wavelengths, output power and response time. An IR sensor consists of an IR LED and an IR photodiode; together they are called a PhotoCoupler or OptoCoupler.

IR Transmitter or IR LED: This is a light-emitting diode (LED) which emits infrared radiations called IR LEDs. The radiation emitted by IR LED is invisible to the human eye and looks like a normal LED (see Figure 9.2) [9].

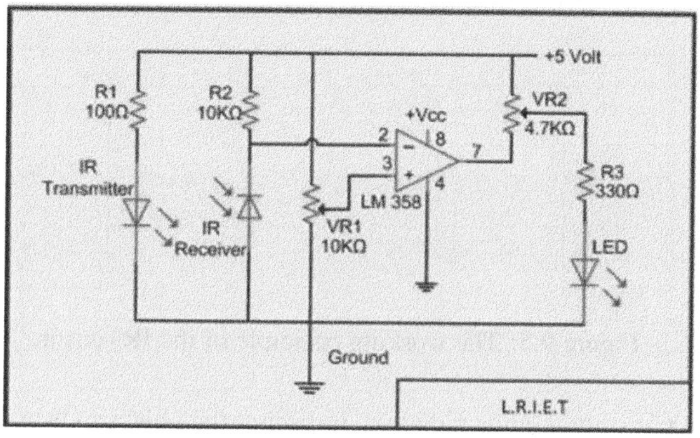

Figure 9.2: Circuit diagram of IR sensor.

An illustration of an infrared LED is shown in Figure 9.3 below.

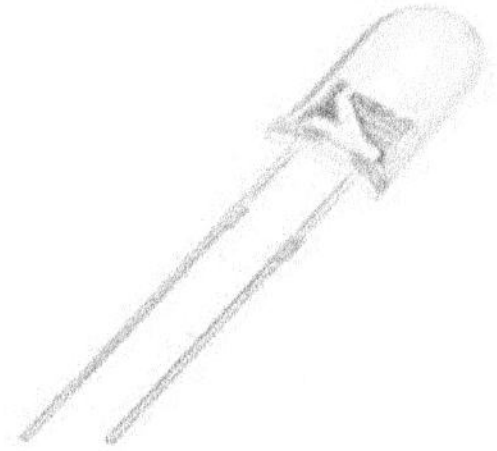

Figure 9.3: The infrared LED.

IR Receiver of Photodiode: Infrared receivers detect the radiation from an IR transmitter. The IR receivers come in the form of photodiodes and phototransistors. Infrared photodiodes detect only infrared radiation, not any other. Figure 9.4 shows an illustration of an IR receiver or a photodiode [9].

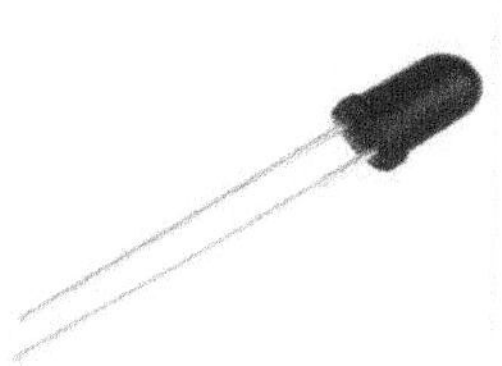

Figure 9.4: The IR receiver or photodiode.

There are several types of IRs which can be differentiated on the basis of the wavelength, voltage, package, etc. When used in an infrared transmitter-receiver combination, the wavelength of the receiver should match with that of the transmitter.

The emitter is an IR LED and the detector is an IR photodiode. The IR photodiode is sensitive to the IR light emitted by an IR LED. The photodiode's resistance and output voltage change in proportion to the IR light received. This is the underlying working principle of the IR sensor [9] (see Figure 9.5).

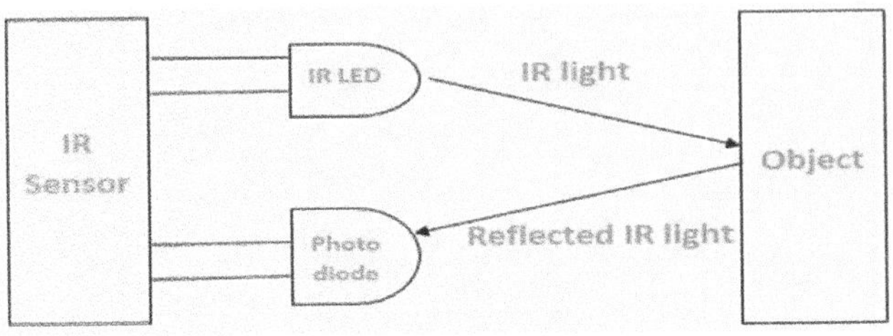

Figure 9.5: The working principle of the IR sensor.

When the IR transmitter emits radiation, it reaches the object and some of the radiation reflects back to the IR receiver. The output of the sensor is defined by the intensity of the reception by the IR receiver.

9.4 IoT Sensors and Board

A lot of sensors are available in IoT technology, some of which are shown below along with there classification and uses.

Figure 9.6 lists the various sensors used in IoT. The most commonly used sensors are listed below in detail.

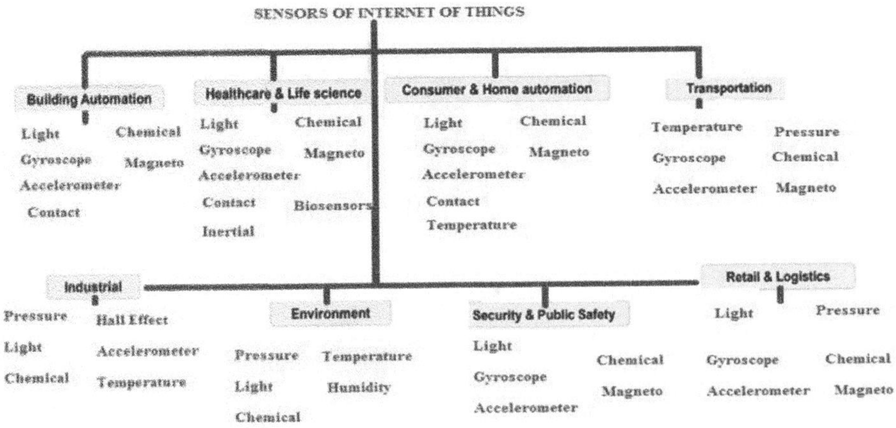

Figure 9.6: Sensors used in IoT.

9.4.1 Arduino UNO Board

An Arduino UNO board is an open-source electronics platform which is capable of using hardware and software. The board is able to read the input from different sensors and produce the output according to the program code [10-13]. We can control the board or enable it to perform any task as per the instructions. There is a set of instructions uploaded on the board in the form of a sketch.

Figure 9.7 shows the Arduino UNO board which is generally used in many IoT projects.

Figure 9.7: Arduino UNO board.

The Arduino board consists of 6 analog pins and 14 digital pins. The board also has other pins, such as (5/3.3) V, which is used to supply current to the attached components; and GND, which works as ground, Rx, Tx, etc. It has 32k of flash memory to store the sketch. It also consists of a reset button to reset the board.

9.4.2 Arduino USB Cable

To upload any sketch on the board we have to connect the board with our system. This connection can be established using the Arduino USB cable shown in Figure 9.8 [10,11].

Figure 9.8: Arduino USB cable.

This cable can also be used to provide the power supply to the board by computer.

9.4.3 Pulse Rate Sensor

The pulse rate sensor is a well-designed plug-and-play heart-rate for Arduino. To detect live heart-rate data the sensor can be clipped onto the fingertip or earlobe and plugged right into the Arduino. It also includes an open-source monitoring app that graphs your pulse in real time. It consists of a 24-inch color-coded cable with (male) header connectors, which makes it easy to embed the sensor and connect to an Arduino [8].

There are Velcro dots on the "hook" side which are also perfectly sized to the sensor; these Velcro dots are very useful for making a Velcro (or fabric) strap that wraps around. Transparent stickers are used on the front of the pulse sensor to protect it from oily fingers and sweaty earlobes. The pulse sensor contains 3 holes around the outside edge, which make it easy to sew into almost anything (see Figure 9.9).

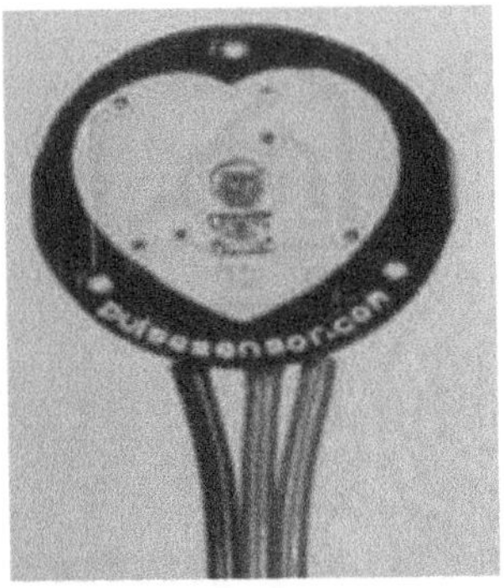

Figure 9.9: Pulse rate sensor.

9.4.4 IR Sensor

Infrared technology addresses a wide variety of wireless applications. In the electromagnetic spectrum, the infrared portion is divided into three regions according to the wavelengths of these regions as specified [14,15] below:

1. Infrared region — 700 nm to 1400 nm
2. Mid infrared region — 1400 nm to 3000 nm
3. Far infrared region — 3000 nm to 1 mm

Infrared sensors are of two types: passive and active. The photo in Figure 9.10 shows the physical appearance of the IR sensor [12,13].

Figure 9.10: IR sensor.

Generally, an infrared sensor is a device which consists of two parts: an emitter, to emit IR rays, and a receiver, to collect the reflected rays. When an object intercepts these rays, then the object is detected. This information is then sent to the Arduino, which further processes it into digital output.

9.4.5 Temperature Sensors

These are one of the commonly used sensors that measure the temperature or heat of a given medium. These sensors use a number of methods to determine and quantify the temperature of any object. Some of the temperature sensors require a physical contact with the object while others do not require contact as they can detect liquid or gases that emit radiant energy like a spike in heat or temperature. Highly sensitive semiconductors available on the market are capable enough to monitor and display slight variations in temperature (see Figure 9.11).

Figure 9.11: Temperature sensor.

9.4.6 Proximity Sensors

Proximity sensors are the best at detecting any type of motion. They are widely used in applications such as security, safety, or efficiency. These sensors are used to avoid obstacles when navigating a crowded place or any complex route and are the best possible sensor for map building. Proximity sensors use electromagnetic radiation like radar signals to detect motion or habitation. They are best used in many types of industry [10,14]. Retailers use proximity sensors to detect the motion between a customer and a product in which he or she is interested in order to send them special offers on their IoT devices. They also can be used in parking systems, museums, airports, etc. (see Figure 9.12).

Figure 9.12: Proximity sensor.

9.4.7 Pressure Sensors

Pressure sensors are used for measuring the pressure of any type of gas or liquid. Pressure sensors convert physical power into an electrical signal. They can also be effectively used for measuring other variables like speed and altitude or other similar situations. Barometers and pressure gauges are the pressure sensors commonly used in an IoT system. Barometers are helpful in weather forecasting as they measure ambient air accurately. Pressure gauges are mostly used in industrial sites as they are good for monitoring pressure in closed environments. Pressure sensors are the ultimate solution for IoT devices as they can be used in various areas such as touchscreen devices, biomedical devices, automotive systems and in the manufacturing industry. Micro pressure sensors are a type of small-size sensors for the measurement of pressure. The first micro sensor was developed and used by industry in a piezoresistive pressure sensor to reduce fuel consumption by maintaining a tight control ratio between air and fuel and another is a disposable blood-pressure sensor to monitor the corresponding status of the patient during operations. The available products on the market are usually either piezoresistive or capacitive. Micro pressure sensors work on the principle of mechanical bending of thin silicon diaphragm by the contact air or gas pressure. This physical movement is converted into electrical output (see Figure 9.13).

Figure 9.13: Pressure sensor.

9.4.8 LCD Display

An LCD liquid crystal display (LCD) screen is an electronic display module that finds a wide range of applications. The reason behind this is that LCDs are economical, easily programmable, have no limitations on displaying special and even custom characters, animation and so on.

Figure 9.14: LCD display.

The LCD display is shown in Figure 9.14 [12,13,16]. The operating voltage of this LCD is between 4.7V to 5.3V. It includes two rows where each row can produce 16 characters. Its size is 16×2. The utilization of current is 1 mA with no backlight.

9.4.9 Relay

A relay is an electrically operated switch that can be used to control the power supply of the flow. It can be controlled by voltage as low as 5V provided by the Arduino pins [17].

Figure 9.15: Relay.

As shown in the Figure 9.15, this relay module has one channel (blue cubes). There are other models with two, four and eight channels. One side of the relay module is connected to the Arduino board to control the relay module and another side is connected with the main voltage or main supply that we want to control.

9.4.10 Power Supply

After uploading the sketch on the board, the board is ready to perform the task for which the code is written but to do this we have to provide a power supply to the board. Any Arduino board has two options for the power supply: using a USB cable and using an external supply.

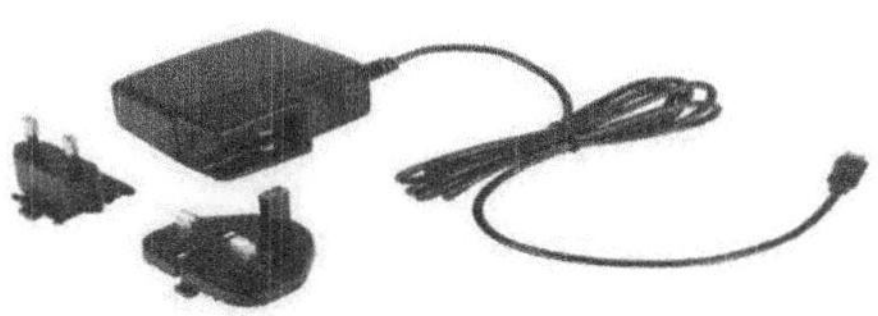

Figure 9.16: Power supply.

Figure 9.16 shows the external power supply which is used to provide the power to the Arduino board without the use of a computer system.

9.4.11 Jumping Wires

A jumping wire is an electrical wire, or set of wires in a bunch, with a pin at each end which is normally used to interconnect the components, internally or with other equipment or components, without soldering [18-26].

There are three types of jumping wires:

- Male to Male jumping wire.
- Female to Female jumping wire.
- Female to Male or Male to Female jumping wire.

Figure 9.17 show all three types of wires. The types of the wires are decided according to the end pins.

Figure 9.17: Jumping wires.

9.5 Use of IoT in COVID-19 Prevention

By analyzing the guidelines provided by WHO for the prevention of COVID-19, we focus on touching the surface of an electric board and switches for various appliances.

The concept of the IoT is to make everything automatic so that it can be controlled remotely and without touching it. The IoT has some properties which are very useful in the prevention of COVID-19 as well as in stopping the spread of the virus.

As seen in the WHO guidelines, the main reason for the spread of the virus is touching. By adhering to the touching concept than the chances and percentage of the virus spreading will decrease.

Some benefits of using the IoT to stop the spread of COVID-19 are:

1. It works on touchless technology.
2. It is an automatic concept.
3. IoT can remotely control any machine.
4. There is no need for sanitization in touchless technology.
5. There is no risk of electric shocks.

The accompanying table compares the risk analysis of current technology with our technology, and this graph makes it simple to see that the COVID-19 guidelines are being used by this board.

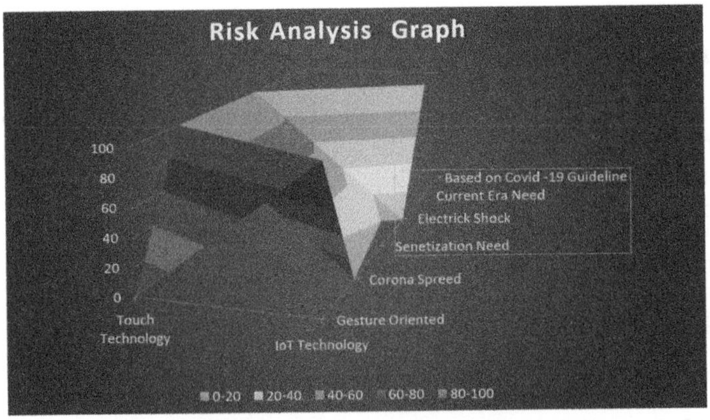

Figure 9.18: Risk analysis.

Virus expansion is very low or we can say that it is approximately 0%. And it may also have other benefits.

9.6 Conclusion

An Arduino circuit board can be used anywhere, including the home, office, school, university, etc. One more advantage of this board is that it can be used with any electronic equipment as it is not bounded. It has an electronic socket which can be used for various appliances. It is also portable and not fixed in the board, thus it can be taken anywhere while traveling. This switch can also be used by kids as it is touchless so there is no danger of electric shock.

This is the latest technology and as it is touchless there is no need to periodically sanitize the switch. The main utility of this concept is to prevent the spread of COVID-19 by using electronic equipment generally used in the same area by various people.

There are some basic advantages of the switch:

- It is portable so it can be used anywhere.
- It can operate any electric appliance according to the relay capacity.
- It is secure from coronavirus spread.
- There is no risk of electric shock.
- It does not harm children.
- No sanitation is needed.

Table 9.5: Comparison of non-IoT and IoT.

Feature	Non-IoT	IoT
Electrick Shock	100	0
Coronavirus Spread	100	0
Use of Sanitization	100	0
User Friendly	50	100
Touching Issue	100	0
Current Need	0	100
Future Scope	0	100

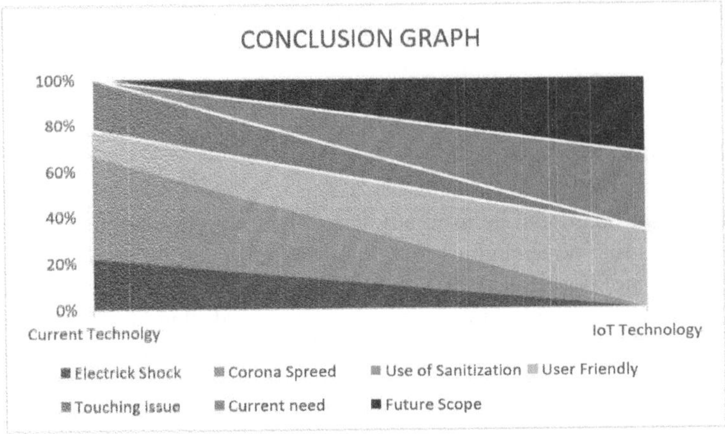

Figure 9.19: The comparison of non-IoT and IoT.

References

1. Hussen, H., & Alemu, Z. A. (2021). Risk of COVID-19 infection and associated factors among healthcare workers: a cross-sectional study at Eka Kotebe Treatment Center in Ethiopia. *International Journal of General Medicine*, 14, 1763.
2. https://www.who.int/
3. Jeppesen, S., & Miklian, J. (2020, May). Introduction: Research in the time of COVID-19. In *Forum for Development Studies* (Vol. 47, No. 2, pp. 207-217). Routledge. DOI: 10.1080/08039410.2020.1780714
4. Cao, X. (2020). COVID-19: immunopathology and its implications for therapy. *Nature Reviews Immunology*, 20(5), 269-270. DOI: 10.1038/s41577-020-0308-3
5. https://randomnerdtutorials.com/
6. Velavan, T. P., & Meyer, C. G. (2020). The COVID-19 epidemic. *Tropical Medicine & International Health*, 25(3), 278. DOI: 10.1111/tmi.13383
7. Fauci, A. S., Lane, H. C., & Redfield, R. R. (2020). Covid-19—navigating the uncharted. *New England Journal of Medicine*, 382(13), 1268-1269.
8. Abbas, J., Wang, D., Su, Z., & Ziapour, A. (2021). The role of social media in the advent of COVID-19 pandemic: crisis management, mental health challenges and implications. *Risk Management and Healthcare Policy*, 14, 1917.
9. Lakshmana Kumar, V. N., Satyanarayana, M., Singh, S., & Le, D. N. (2022). A Novel Compact Frequency and Polarization Reconfigurable Slot Antenna Using PIN Diodes for Cognitive Radio Applications. In *Smart Antennas* (pp. 85-95). Springer, Cham.
10. Jassim, G., Jameel, M., Brennan, E., Yusuf, M., Hasan, N., & Alwatani, Y. (2021). Psychological impact of COVID-19, isolation, and quarantine: A cross-sectional study. *Neuropsychiatric Disease and Treatment*, 17, 1413.
11. https://www.electronicshub.org
12. SaravanaKumar, U., Madhumitha, K., & Manikandan, S. N. (2021). Automatic Room Light Controller with Bidirectional Visitor Counter. *International Scientific Journal of Contemporary Research in Engineering Science and Management*, 5(4), 153-160.

13. Waradkar, G., Ramina, H., Maitry, V., Ansurkar, T., Rawat, M. A., & Das, M. P. (2016). Automated room light controller with visitor counter. *Imperial Journal of Interdisciplinary Research*, 2(4), 777-780. DOI 10.4010/2016.638.

14. Louis, L. (2016). Working principle of Arduino and using It as a tool for study. *International Journal of Control, Automation, Control, Communications and Systems*, 1.

15. Hossain, M. S., & Nahiyan, H. (2014, December). Automatic Control System for Lighting of a Single Door Room with Bidirectional People Counter. In *International Conference on Mechanical, Industrial and Energy Engineering* (pp. 26-27).

16. https://www.arduino.cc

17. https://www.elprocus.com

18. Le, T. T., Andreadakis, Z., Kumar, A., Román, R. G., Tollefsen, S., Saville, M., & Mayhew, S. (2020). The COVID-19 vaccine development landscape. *Nat Rev Drug Discov*, 19(5), 305-306. DOI: 10.1038/d41573-020-00073-5

19. Al-Waisy, A. S., Mohammed, M. A., Al-Fahdawi, S., Maashi, M. S., Garcia-Zapirain, B., Abdulkareem, K. H., ... & Le, D. N. (2021). COVID-DeepNet: hybrid multimodal deep learning system for improving COVID-19 pneumonia detection in chest X-ray images. *Computers, Materials and Continua*, 67(2), 2409-2429.

20. Saxena, A., Taterh, S., & Saxena, N. (2020). Comparative Study of the Ultrasonic and Infrared Person Counter. In *Soft Computing: Theories and Applications* (pp. 1081-1092). Springer, Singapore.

21. https://www.mohfw.gov.in/pdf/PreventivemeasuresDOPT.pdf

22. Le, D. N., Parvathy, V. S., Gupta, D., Khanna, A., Rodrigues, J. J., & Shankar, K. (2021). IoT enabled depthwise separable convolution neural network with deep support vector machine for COVID-19 diagnosis and classification. *International Journal of Machine Learning and Cybernetics*, 12(11), 3235-3248.

23. https://ourworldindata.org/coronavirus/country/india

24. Shermi, S. C. S. S. D., Soundarya, S., & Manickavasagam, M. R. (2020). SMART MATERNAL REAL TIME MONITORING USING IoT-TECHNIQUE. *International Research Journal of Engineering and Technology*, 7(3).

25. Krishnamurthi, R., Kumar, A., Gopinathan, D., Nayyar, A., & Qureshi, B. (2020). An overview of IoT sensor data processing, fusion, and analysis techniques. Sensors, 20(21), 6076. DOI: 10.3390/s20216076

26. Prasad Sharma, K., Walia, K., & Gupta, S. (2023). IoT for fight against COVID-19. In *Next Generation of Internet of Things* (pp. 585-596). Springer, Singapore.

10

Role of Satellites in Agriculture

Prashant Johri[1], Kanta Prasad Sharma[2], Aadrit Chauhan[1], Sunilkkhatri[1]

[1] School of Computer Science and Engineering, Galgotias University Uttar Pradesh, India
[2] Computer Engineering and Application , GLA University Mathura, India
Email: singhjn2000@gmail.com, johri.prashant@gmail.com, proudygarv@gmail.com, tokpsharma@gmail.com

Abstract

There has been an immense amount of development in the field of agriculture during the last two decades and one of the major areas of technology which have made it possible is remote sensing technologies. Satellite imagery has proved to be a boon for precision agriculture and has increased production efficiency as well as aided farmers with the help of machine learning and big data to make the right decision at the right time. This chapter reviews the role of satellites in agriculture and how big data analysis can give amazing results, hence contributing to the national economy. Furthermore, the advantages and disadvantages of these technologies along with the challenges that lie ahead are discussed and how the future of these technologies will help the agricultural sector.

Keywords: Remote sensing, agriculture, big data, machine learning

10.1 Introduction

The agriculture sector in India contributes about 16% to the GDP and is the third largest contributor after the Services and Production sector. India is investing in the latest technologies to come up with new methods to increase the revenue of this sector.

One of the major technologies that is being widely used is the data from satellites using multispectral and hyperspectral imaging [1]. The data from the satellites have been helping in various fields of agriculture, viz. estimating the time of harvest, predicting in-season yields, detecting and controlling pests and diseases, understanding water and nutrient status, planning crop nutrition programs and taking decisions about in-season irrigations, etc. The above-mentioned techniques and processes are not determined by directly looking at the images and the data. Instead, an in-depth analysis must be done, which involves a lot of techniques and processes.

After applying the processing techniques, the soil and crop conditions can be determined through leaf area analysis, mineral analysis, climate analysis, temperature variations, water quality, etc. Based on the results obtained, further analysis is carried out and trends are realized which help make decisions on irrigation requirements, fertilizer requirements, mixing of different crops to give optimum yield, forecasting the perfect time for harvesting and cultivation, etc. These decisions, when applied to a particular area or patch of land, are called precision agriculture.

Precision agriculture uses the latest technologies driven by satellites and remote sensing which support precision agriculture. Precision agriculture focuses on a specific area rather than considering the whole area since the texture, soil, humus, temperature, moisture content, etc., are not uniform throughout the same field; therefore, precision agriculture uses the approach of working on those parameters area wise rather than as a whole; for example, using fertilizer only where it is required rather than spraying it all over the field (see Table 10.1).

Table 10.1: Comparison of satellite types.

Satellite Type	Description
Open Data Satellites	Data from these satellites are unclassified i.e. they are freely available for everyone to use.
Commercial Satellites	These satellites are used for civilian, government or non-profit use, they are not used for military or any human space flight program.
Weather Satellites	These satellites are used for collection information that is used for weather forecasting.
Geodesy Satellites	These satellites are used to measure the form and dimensions of Earth, the location of objects on its surface, its variations, etc.
Ocean Satellites	These satellites are used to study oceans and can be used to survey oceanic life.

This not only helps in reducing the overall cost, but also using the inputs wherever necessary helps to optimize harvest growth; hence, helps in making accurate decisions in a timely manner. Decision-making plays an important role in the agricultural sector as the time to cultivate, spray pesticides, fertilizers, etc., are to be decided accurately and untimely inputs can make all the efforts in vain. The emerging advancements that encompass data visualization, processing, and in-depth analysis aid the person on the ground in real time in contrast to the decision being made solely on the experience and knowledge of farmers.

These technologies bless us with great benefits, viz. reducing cost, maximizing efficiency, optimizing use of inputs (like seeds, insecticides, pesticides, etc.), determining the correct amount of inputs needed; hence increasing the overall efficiency. This will help alleviate the huge challenges facing farmers as the climate and overall environment continue to drastically change.

10.2 Processing Satellite Images

The big data from the satellites is is not ready for use due to its complexity and the inability of a normal person to understand it. The images from the satellites are of very large size and hence require computers with high computational and processing capabilities. Apart from the high system requirements, the images must be processed before analyzing them (see Figure 10.1).

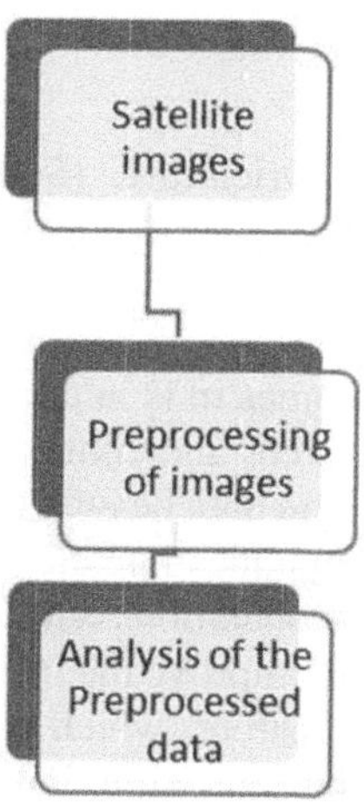

Figure 10.1: Data processing.

Every object has the ability to absorb, reflect and transmit waves, and this property of the objects is exploited and used in the remote sensing technology. These interactions of objects are then analyzed in the microwave, infrared and visible light regions.

According to [2], the remote sensing systems include visible NIR (near-infrared) (0.4–1.5 mm) sensors for plant vegetation studies, SWIR (short wavelength infrared) (1.5–3 mm) sensors for plant moisture studies, TI (thermal infrared) (3–15 mm) sensors for crop field surface or crop canopy temperature studies, and microwave sensors for soil moisture studies.

The programs of MODIS (moderate resolution imaging spectroradiometer) (National Aeronautics and Space Administration [NASA], Washington DC), Landsat (NASA and United States Geological Survey [USGS], Reston, VA), European satellites such as SPOT (SPOT Image, Toulouse, France), and Chinese satellites such as Ziyuan (China Center for Resources Satellite Data and Application, Beijing, China) all provide products at different levels more or less depending on different applications [2] (see Figure 10.2).

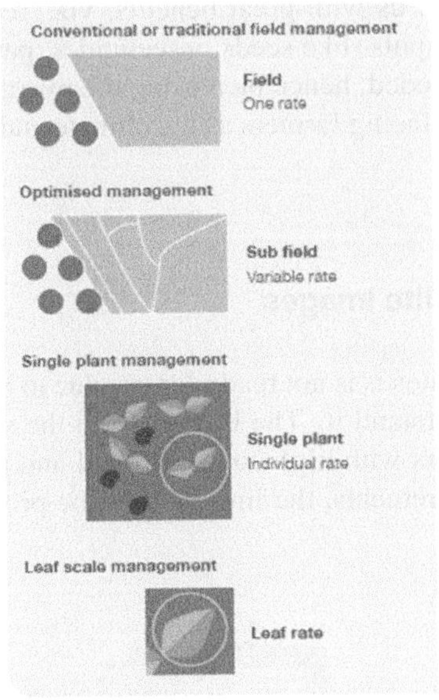

Figure 10.2: Example of precision agriculture. (Source: Satellite Applications Catapult, UK)

Pre-processing: The data that comes in is in a raw form and has a lot of irregularities, and the sampling is not very consistent and contains speckles and various other anomalies. These irregularities can affect the data heavily and the result hinders accurate results. Pre-processing includes several techniques like cleaning, integration, transformation, and reduction. It is a common name for operations with images at the lowest level of abstraction. The aim of pre-processing is to improve the image data that suppresses the unwanted distortions and enhances the image features which are important for further processing.

Recent studies [3,4] showed how precision agriculture may benefit from processing imagery. Satellite imagery datasets became readily available by open access to NASA Landsat in 2008 [5] and to ESA Sentinel satellites.

10.3 Product Levels of Satellite Remote Sensing Data

The normalized difference vegetation index (NDVI) is one of the attributes that is being used to process the data into a usable form to predict the concentration of chlorophyll or biomass, which can later be interpreted in many ways to determine various other factors.

10.3.1 A Brief Discussion and Review of Analysis Techniques

There are a lot of techniques that are being deployed for the purpose of analyzing images and the most popular among them are deep learning (DL), machine learning (ML), ($k-$means, support vector machines (SVM), artificial neural networks (ANN), linear polarizations, wavelet-based filtering, vegetation indices (NDVI) and regression analysis [6-18] (see Figure 10.3).

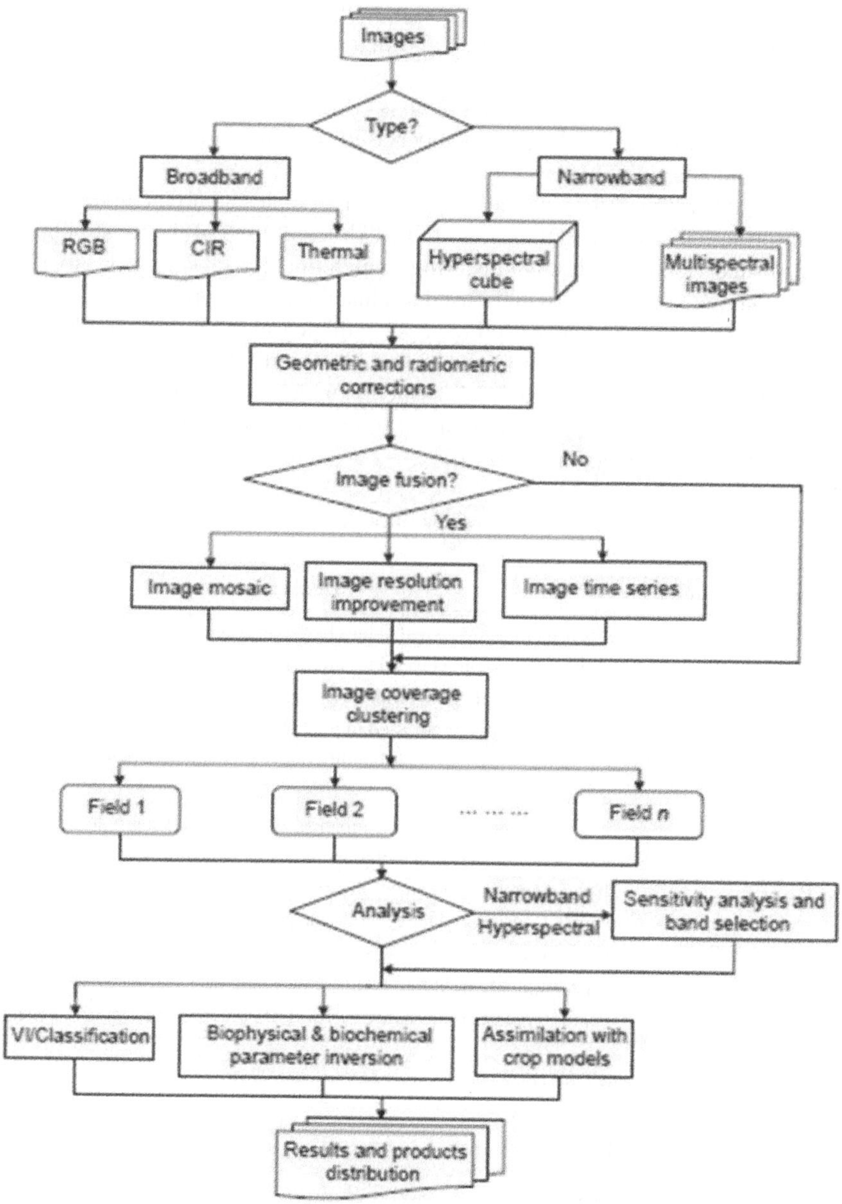

Figure 10.3: Remote sensing image processing, analysis and management flow.

10.3.2 Machine Learning

Machine learning is the scientific field that gives the ability to learn without being strictly programmed [8]. The machine learning models are trained using training data and then this trained model is used to analyze the new data that has to be tested. Based on the type of learning, the machine learning tasks can be classified into two categories: supervised and unsupervised learning.

Supervised learning: Here, both input and output data sets are available and the functional relationships between the two are to be determined. This relation is achieved by using regression techniques (the most popular are linear regression, logistic regression, and stepwise regression). Also, Bayesian models (BM), which basically involve probability, are used for accomplishing the task of supervised learning and determining the relation between the input and output.

Unsupervised learning: Here, we only have the input data set and the underlying pattern or useful information from the data set has to be extracted. One of the most popular applications of unsupervised learning is clustering, which enables grouping of closely related data. The technique that is most widely used to cluster data is $k-$means.

Other models which are widely used to analyze the learning are instance-based models (IBM), decision trees, artificial neural networks (ANN), etc. The deep ANNs or the deep learning is an extremely potent area of ML which is widely referred to in a lot of journals. It has gained its popularity in a short amount of time and has emerged as the pioneer in the field of big data analysis (see Table 10.2).

Table 10.2: Comparison of machine learning algorithm types.

Types of Algorithms	Description
Linear Regression	It is used to estimate real values based on continuous variables such as cost of house, upcoming taxes, etc. Eq: $Yax + b$.
Logistic Regression	It is a classification algorithm used to assume discrete values, i.e., 0 or 1, true or false, yes or no (used to predict events which have only 2 results). Eq: $odds = \frac{p}{(1-p)}$ = probability of event occurring / probability of event not occurring.
Decision Tree	It is used for classification problems. In this process a problem is divided into as many parts as possible in understandable terms.
Naïve Bayes	It is a theorem based on prediction of the features of a particular class, which is considered unrelated to the presence of any characteristic. Eq: $P(c\|x) = \frac{P(X\|c)P(c}{P(x)}$
kNN (k-Nearest neighbor)	It is commonly used for classification problems. It stores all possible conditions and groups new cases according to the votes by its $k-$neighbors.

10.3.3 Deep Learning

Deep learning, one of the most commonly used techniques, has gained momentum over the past five years in the agricultural sector [9]. Deep learning presents the data in hierarchical form, representing the convoluted deeper parts and making it easy to understand, thereby increasing learning capabilities, performance and accuracy. Deep learning is the more in-depth version of the machine learning model, representing the data in hierarchical form through several levels of abstraction [8,9].

Deep learning has the ability to extract features from hierarchical compositions and can solve problems in less time as compared to the existing techniques. The edge of dl over others is because of the hierarchical composition that can be applied extensively and to various types of data sets like video, audio, etc.

It has gained a lot of popularity particularly in the agricultural field for the above-mentioned reasons and its flexibility (see Figure 10.4).

Figure 10.4: Online farm management platform that exploits computer vision for crops.

Many research papers have shown the immense ability of ML to analyze data, such as the time taken to do tasks manually, by using experiences which are not always reliable. The approaches of training the models can give amazingly accurate results in less time. According to [10], deep learning takes a longer time to train but the execution and testing time is quite commendable with respect to other techniques. Below are a few cases taken from research papers which show the wide-ranging features used to make predictions based on crop observation.

1. Automatic counting of coffee fruits on a coffee branch using 42 color features in a digital image illustrating coffee fruits [11].

2. Detection of cherry branches with full foliage using color digital images [12].

3. Identifying immature green citrus fruits under natural outdoor conditions using imaging properties like coarseness, regularity, and granularity [13].

4. Estimation of grassland biomass using vegetation indices and spectral bands of red and near-infrared imaging [14].

5. Prediction of wheat yield within-field variation using the normalized values of online predicted soil parameters and satellite NDVI [15].

6. Rice development stage prediction and yield prediction using features like surface weather analysis and soil physicochemical data with yield or development [16].

Also, diseases can be predicted using the above-mentioned features. There are papers on weed detection too. The quality of crops has also been improved using the features stated above (see Figures 10.5 and 10.6) [17].

Table 10.3: Comparison of deep learning algorithm types.

Types of Algorithms	Description
Backpropagation	It is used for training feedforward neural networks for controlled learning; it is also responsible for solving an expression for the cost derivative function.
Feedforward Neural Networks	These are usually fully interconnected, which means that every neuron in a layer is connected to every other neuron in the other layers.
Convolution Neural Network	It is a method of combining two functions by multiplying them for mathematical reasons; it calculates how many multiple functions correspond with each other as they pass over each other.
AutoEncoders	These are neural networks that are distributed directly and they also increase the input strength at the output; they also have a layer of hidden code that explains the model.
Generative Adversarial Networks	These are the upcoming popular ML learning model for e-commerce because of their skill in receiving, understanding and reconstructing visuals.

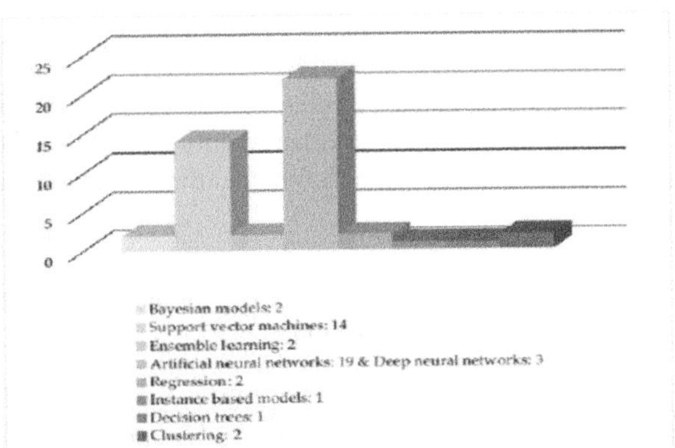

Figure 10.5: ML models with their total rate (observed in total 40 papers).

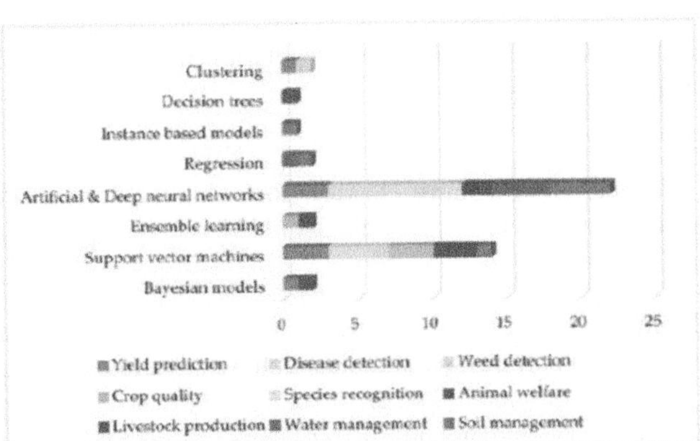

Figure 10.6: Total number of ML models in each subcategory of four main categories.

The above models are largely employed in predicting weather conditions, and further, determine many of their other associated features which help immensely in the prediction of a lot of processes and decisions in agriculture. Here are a few cases:

1. Determining the soil texture, moisture content, soil temperature, etc., using technologies like ANN, SVM, IBM, KNN, etc.
2. Prediction of water bodies using satellite data (the results are more pronounced in the case of large water bodies) and monitoring the quality of water using regression techniques, ANN based scenario generation, ELM/GRNN, MARS, etc.
3. Monitoring air content and dust using data from MODIS, Sentinel, etc.

10.4 Future Challenges

There are a number of challenges posed as well as we seamlessly flow across into the new emerging technologies.

The data has to be reliable enough to make sure the efforts and capital do not go in vain. This can only be ensured if the data from certified and government-regulated satellites are used and are made available readily. The agricultural-related problems include seed identification, soil and leaf nitrogen content, irrigation, water erosion assessment, pest detection, herbicide use, identification of contaminants, diseases or defects on food, crop-hail damage, and greenhouse effect. The monitoring and data analysis techniques applied to address these is similar to deep learning according to [18-20]. Furthermore, these technologies have to be applied in livestock management as well at a much larger scale than now (see Table 10.4).

Table 10.4: The tools website.

Name	Website	Free Access
USGS Earth Explorer	https://www.usgs.gov/	YES
Sentinel Open Access Viewer	https://scihub.copernicus.eu/	NO
NASA Earthdata Research	https://search.earthdata.nasa.gov/search	YES
NOAA Data Access Viewer	https://coast.noaa.gov/dataviewer/#/	NO
Digital Globe Open Data Program	https://www.digitalglobe.com/ecosystem/open-data	YES
GEO Airbus Defense	http://www.intelligence-airbusds.com/	YES
NASA Worldview	https://worldview.earthdata.nasa.gov/	NO
NOAA CLASS	https://www.bou.class.noaa.gov/saa/products/	YES
National Institute for Space Research	https://www.natureindex.com/	YES
Bhuvan Indian Geo-Platform of ISRO	https://bhuvan.nrsc.gov.in/bhuvan_links.php	NO
JAXA's Global ALOS 3D World	https://www.eorc.jaxa.jp/ALOS/en/aw3d30/index.htm	NO
VITO Vision	https://www.vito-eodata.be/PDF/portal/Application.html	NO
NOAA Digital Coast	https://coast.noaa.gov/digitalcoast/	NO
Satellite Land Cover	https://www.isro.gov.in/	NO
UNAVCO	https://www.unavco.org/	NO

The big data should be made available at a rapid rate and for this purpose, proper agricultural data management organization should come into play with the aid and cooperation of both private and government organizations. Also, the security of this data also poses a challenge. The management and organization of data should be done at different levels, such as village, local, national and global levels, so that the data will be uniform and consistent.

10.5 Conclusion

There is an immense ocean of information yet to be explored on the role of satellites in agriculture, which has been a blessing in the agricultural field and has given rise to precision agriculture. The acquisition of resources still needs to be standardized and the network speed at which data can be uploaded and downloaded still needs to be improved. Real-time data feed should be made available and platforms should be available with high computational capabilities that could process the big data rapidly and in less time with efficiency. Videos from the satellites should be made in real time so that the monitoring can also be done in real time. The communication media between the satellites and the various other sensors and systems should be efficient and optimized and their synchronization will give great results. This will also lead to competition among organizations, which will eventually lead to better outcomes in the near future (see Table 10.5).

Table 10.5: Comparison of communication media between satellites.

Satellite	Description
RADARSAT	To facilitate the management of resources (including farmland), marine surveillance, ecosystem monitoring, ice monitoring, disaster management and mapping in Canada and around the world.
SMAP	To map soil moisture and freeze/thaw status.
SMOS	To map sea surface salinity and monitor soil moisture on a global scale, thus contributing to a better understanding of the Earth's water cycle.

References

1. Ishimwe, R., Abutaleb, K., & Ahmed, F. (2014). Applications of thermal imaging in agriculture—A review. *Advances in remote Sensing*, 3(03), 128.
2. Huang, Y., Chen, Z. X., Tao, Y. U., Huang, X. Z., & Gu, X. F. (2018). Agricultural remote sensing big data: Management and applications. *Journal of Integrative Agriculture*, 17(9), 1915-1931.
3. Jeppesen, J. H., Jacobsen, R. H., Jørgensen, R. N., Halberg, A., & Toftegaard, T. S. (2017). Identification of high-variation fields based on open satellite imagery. *Advances in Animal Biosciences*, 8(2), 388-393.
4. Jeppesen, J. H., Jacobsen, R. H., Jørgensen, R. N., & Toftegaard, T. S. (2022). Towards data-driven precision agriculture using open data and open source software. *arXiv preprint arXiv:2204.05582*.
5. Wulder, M. A., Masek, J. G., Cohen, W. B., Loveland, T. R., & Woodcock, C. E. (2012). Opening the archive: How free data has enabled the science and monitoring promise of Landsat. *Remote Sensing of Environment*, 122, 2-10.
6. Saxena, L., & Armstrong, L. (2014). A survey of image processing techniques for agriculture. *Proceedings of Asian Federation for Information Technology in Agriculture, Australian Society of Information and Communication Technologies in Agriculture*. Perth, Australia.

7. Huang, J., Tian, L., Liang, S., Ma, H., Becker-Reshef, I., Huang, Y., ... & Wu, W. (2015). Improving winter wheat yield estimation by assimilation of the leaf area index from Landsat TM and MODIS data into the WOFOST model. *Agricultural and Forest Meteorology*, 204, 106-121.

8. Samuel, A. L. (2000). Some studies in machine learning using the game of checkers. *IBM Journal of Research and Development*, 44(1.2), 206-226.

9. Lee, S. H., Chan, C. S., Wilkin, P., & Remagnino, P. (2015, September). Deep-plant: Plant identification with convolutional neural networks. In *2015 IEEE international conference on image processing (ICIP)* (pp. 452-456). IEEE.

10. Christiansen, P., Nielsen, L. N., Steen, K. A., Jørgensen, R. N., & Karstoft, H. (2016). DeepAnomaly: Combining background subtraction and deep learning for detecting obstacles and anomalies in an agricultural field. *Sensors*, 16(11), 1904.

11. Ramos, P. J., Prieto, F. A., Montoya, E. C., & Oliveros, C. E. (2017). Automatic fruit count on coffee branches using computer vision. *Computers and Electronics in Agriculture*, 137, 9-22.

12. Amatya, S., Karkee, M., Gongal, A., Zhang, Q., & Whiting, M. D. (2016). Detection of cherry tree branches with full foliage in planar architecture for automated sweet-cherry harvesting. *Biosystems Engineering*, 146, 3-15.

13. Sengupta, S., & Lee, W. S. (2014). Identification and determination of the number of immature green citrus fruit in a canopy under different ambient light conditions. *Biosystems Engineering*, 117, 51-61.

14. Ali, I., Cawkwell, F., Dwyer, E., & Green, S. (2016). Modeling managed grassland biomass estimation by using multitemporal remote sensing data - A machine learning approach. *IEEE Journal of Selected Topics in Applied Earth Observations and Remote Sensing*, 10(7), 3254-3264.

15. Pantazi, X. E., Moshou, D., Alexandridis, T., Whetton, R. L., & Mouazen, A. M. (2016). Wheat yield prediction using machine learning and advanced sensing techniques. *Computers and Electronics in Agriculture*, 121, 57-65.

16. Su, Y. X., Xu, H., & Yan, L. J. (2017). Support vector machine-based open crop model (SBOCM): Case of rice production in China. *Saudi Journal of Biological Sciences*, 24(3), 537-547.

17. Liakos, K. G., Busato, P., Moshou, D., Pearson, S., & Bochtis, D. (2018). Machine learning in agriculture: A review. *Sensors*, 18(8), 2674.

18. Singh, A., Ganapathysubramanian, B., Singh, A. K., & Sarkar, S. (2016). Machine learning for high-throughput stress phenotyping in plants. *Trends in Plant Science*, 21(2), 110-124.

19. Dalal, S., Jaglan, V., & Le, D. N. (Eds.). (2021). *Green Internet of Things for Smart Cities: Concepts, Implications, and Challenges*. CRC Press.

20. Sharma, K. P., Poonia, R. C., & Sunda, S. (2018, August). Accurate real-time location map matching algorithm for large scale trajectory data. In *2018 7th International Conference on Reliability, Infocom Technologies and Optimization (Trends and Future Directions)(ICRITO)* (pp. 646-651). IEEE

11

Search Engine Evaluation Methodology

JN Singh[1], Prashant Johri[1], Gaurav Dhuriya[1], Kanta Prasad Sharma[2]

[1] Galgotias University, Greater Noida, India
[2] Computer Engineerign and Application , GLA University Mathura, India
Email: singhjn2000@gmail.com, johri.prashant@gmail.com, proudygarv@gmail.com, tokpsharma@gmail.com

Abstract

In this chapter, we explore the various methodologies and statistical measures that have been used for better evaluation of search engines. It also includes a comparative study of statistical and automatic methods and the study of various factors that affect the search engine evaluation process. The chapter presents a broad discussion about the various forums that have been used and provides various data for better evaluation of search engines. Thereafter, the popular statistical measures that have been used to evaluate the search engines are discussed. The chapter also details the automatic approaches of click-through data and eye tracking, and provides the reasons as to why the automatic approach for page quality-related judgments collection is considered superior to traditional approaches.

Keywords: Search engine, eye tracking, statistical method

11.1 Introduction

Today, it is a monumental process to judge the performance of information retrieval systems. As we know, some information retrieval systems that satisfy our minimum requirements are better than others. However, it is not an easy task to satisfy the needs and and address the problems of each searcher.

These basic needs are quality content on web pages, information coverage, results ranking and interface. Evaluation is a key technology for making continuous and smooth progress towards constructing a better search engine. In the search engine evaluation process, a search engine's performance is measured in respect to its efficiency and effectiveness. However, the effectiveness of a search engine is measured by how many retrieved documents are relevant to the searcher's needs. In w3 environment, effectiveness is affected by various factors such as the evaluation algorithm used to sort the search results, facilities to provide help for query formulation and techniques used to fix relevance score to results. So with the help of various methods we can measure the engine's effectiveness, which is very important and often necessary. Here, we will focus on the effectiveness of the search engine and less on its efficiency.

11.2 Performance Evaluation Forum

The objective of various evaluation forums is to enhance the performance of information retrieval systems by providing large test collections of structured documents. These forums provide us with an environment in which to test various information retrieval techniques against standard benchmarks [1,2].

11.2.1 Text Retrieval Conference

The main objective of the Text Retrieval Conference (TREC) [3], which is co-sponsored by NIST (US Government), is to support research in all communities involved in researching information retrieval systems by providing the platform essential for large-scale evaluation of text retrieval techniques.

The main purpose of TREC is:

- To motivate research for IR with huge volume of data.
- To increase interaction and communication among various organizations and governments of several countries by creating an open forum for the exchange of research-oriented ideas.
- To speed up new technology transfer into the whole world.
- To encourage the development of new evaluation techniques for information retrieval systems.

11.2.2 Text Retrieval Conference Tracks

Text Retrieval Conferences provide a large test collection of data called TREC tracks, to find new areas for information retrieval research and create the necessary infrastructure for these newly emerged areas using these tracks. The tracks that have been used in new areas of information retrieval research also provide a better environment for evaluation. Every track is assigned some task by various teams of scientists. Evaluation of an information retrieval system is usually done with the help of TREC queries. TREC tracks also help in the research and development of multilingual and multimedia information retrieval.

The main objective of the TREC system is global and parallel testing. It provides an updated document collection to participants.

The Cross-Language Evaluation Forum (CLEF) works on information retrieval for European languages in both monolingual and cross-language contexts.

The forum for Information Retrieval Evaluation (FIRE) is conducted in coordination with the Indian Statistical Institute, Kolkata. The objective of FIRE is to encourage research work in South Asian language information access technologies. Evaluating the performance of search engines is still a big issue in the research area of information retrieval.

On the basis of previous studies, we have categorized evaluation methods into two parts:

(1) Statistical methods

(2) Automatic methods

We can differentiate these two on the basis of page quality assessment. In the statistical approach [4] of evaluation the judgments about page quality are collected with the help of the searchers' visions (see Table 11.1).

Table 11.1: Evaluation methods.

Statistical methods	Automatic method
Precision	Click-Through Data
Recall	Eye Tracking
DCG	
NDCG	

11.3 Search Engine Evaluation Parameters

11.3.1 Precision

Precision [5] is the ratio of the total relevant records retrieved to the total number of irrelevant and relevant records against any query. Precision considers that the probability of randomly selected and retrieved documents is relevant. The search information retrieval system also presents the most relevant results at the top of the ranking system. Precision measures the exactness of retrieving relevant documents in the information retrieval process. It also measures how many documents are relevant in total retrieved web documents. It does not worry if we are not retrieving all the relevant documents but we get penalized if we are retrieving non-relevant documents from the web. A precision score of one means that each record retrieved by a search engine was relevant but not 100 percent true (see Figure 11.1).

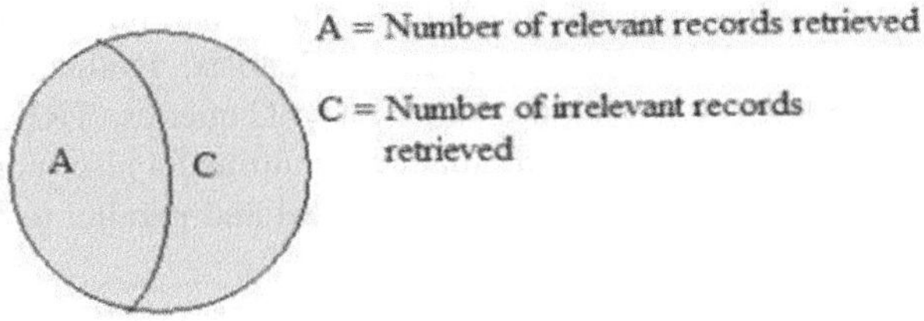

Figure 11.1: Precision

Where A is a collection of relevant records which are retrieved and C is the number of irrelevant records; hence, the precision can be defined as:

$$Precession = \frac{A}{A+C} \times 100 \tag{11.1}$$

or

$$Precession = \frac{NumberoftheRelevantDocumentsRetrieved}{TotalNumberofDocumentsRetrieved} \times 100 \tag{11.2}$$

11.3.2 Recall

Recall [6] refers to the ratio of the total number of relevant records retrieved to the total number of relevant records available in the database.

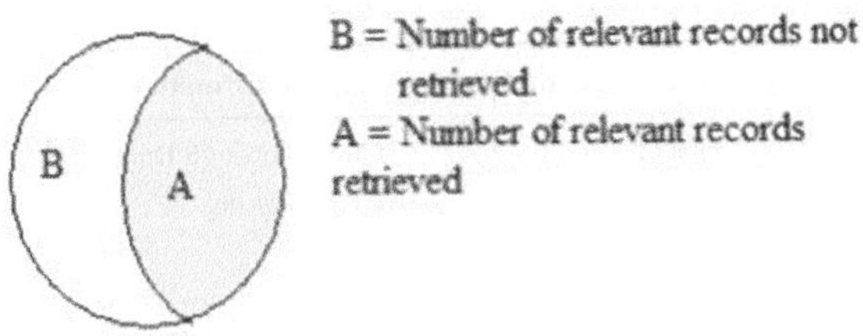

Figure 11.2: Recall

Where B is the total number of relevant documents, so recall can be defined as:

$$Recall = \frac{A}{A+B} \times 100 \tag{11.3}$$

or

$$Recall = \frac{NumberoftheRelevantDocumentsRetrieved}{TotalNumberofRelevantDocuments} \times 100 \tag{11.4}$$

11.3.3 Problems with Precision and Recall

We concluded that precision and recall were the two criteria that have been mostly used to evaluate the performance of information retrieval systems. But it is amazing that these two measures also hold some problems. As noted earlier, object quality must be known in advance in terms of binary judgments, either relevant or irrelevant, during the process of calculating precision and recall. Obviously, few objects can exist which are marginally relevant or somewhat relevant. A few others may be closest in relevance and completely irrelevant in the web. This problem is very complicated and completely depends upon individual perception: what is relevant to one person may not be relevant to others.

Measuring recall is difficult because it is problematic to know how many relevant records exist in a database.

11.3.4 Discounted Cumulative Gain

Discounted cumulative gain (DCG) is a metric that does not use binary decisions about the quality assessment of web pages [7]. Instead, the graded relevance score is assigned to the documents. DCG measures the usefulness or gain for a document based on its position in the results list. DCG may be defined as the sum of the "gain" of showing a specific document multiplied by "discount" of showing it at a particular rank when maximum rank is fixed (Lets say n).

$$DCG = \sum_{\gamma=1}^{n} gain_{\gamma} \times discount_{\gamma} \tag{11.5}$$

"Gain" is defined as the relevance numeric score decided by the searchers and "discount" defines the reciprocal dependency among the ranking.

Two assumptions are made in using DCG and its related measures:

(1) Highly relevant documents become more useful when appearing earlier in a search engine result list (have higher ranks).

(2) Highly relevant documents are more useful than marginally relevant documents.

11.3.5 Normalized DCG

The results positions may vary, if different forms of a single query are submitted to the search engines. Comparing a search engine's performance for a set of queries cannot be consistently achieved using DCG [8] alone, so the cumulative gain at each position P should be normalized for queries. This is done by sorting documents of a result list by relevance scores. The sorting is done in such a way that most relevant results get few top positions in the retrieved citations list. This is called Ideal DCG. For a query, the normalized discounted cumulative gain, or nDCG, is computed as:

$$nDCG_p = \frac{DCG_p}{IDCG_p} \tag{11.6}$$

The DCG values for all queries can be averaged to get a measure of the average performance of a information retrieval systems' ranking algorithm.

The statistical measures for the search engine's evaluation are still used because these measures provide accurate decisions about the page quality. The main disadvantage with the statistical measures is their slow evaluation. The slow evaluation is not suitable for such a large pool of documents because web objects require repeated evaluation for inverted index updating and its maintenance.

Ranking search results effectively is a still unsolved problem in the field of the information retrieval area of research. The existing measures that provide some help in this direction, preliminarily focus on similarity between query terms and keywords of the web page. These measures also consider the whole page quality to compute numeric scores. On the basis of these numeric scores the results are arranged to form the ranked list. In this list, the result having the maximum numeric score is kept on the top while the result that holds minimum numeric score is positioned on the lowest position in the list.

A search engine that updates its inverted index most frequently is considered good. For this updating, quick judgments about the page quality are periodically required, which is not possible manually. For the rapid collection of page quality-related judgments, searchers' interactions with the web browser are examined.

There are two automatic approaches used to collect page quality-related judgments automatically:

(1) Click-through data collection

(2) Eye tracking method

11.3.6 Click-Through

Click-through is considered the most effective technique for collecting searchers' judgments about the usefulness of the web documents. In this technique, it is believed that the total number of clicks-hits on a particular document is directly proportional to the usefulness of web documents. In other words, it is believed that the usefulness of the document increases or decreases with the number of click-hits. In this technique of judgments collection, the searchers are not required to perform additional tasks, the clicks-hits which are naturally hit by searchers are used to calculate the usefulness of the web pages. Searchers remain unaware of the hidden data collection. With the click-through data, the decisions are collected from millions of users from all the geographical areas of the world, so the chances of results being biased are removed. The biggest advantage with the click-through data is its fast assessment nature. It starts the web-page evaluation as quickly as it gets uploaded on the web. This technique of users' judgments collection does not require costly and complicated infrastructure. This work can be implemented with a normal program or through software.

11.3.7 Eye Tracking

Eye tracking approach is also popular for its usability in the area of performance evaluation of search engines. It is an approach by which an individual's eye movements are tracked during the searching process. It is a bit of a difficult process because it requires some scientific instruments which are costly and not easy to handle. With the help of this approach the observer knows where a person is looking at any particular time on the monitor and the sequence of eye shifting from one position to another.

We categorize eye movements' action according to the various indicators of eye-related behaviors like eye fixations and its position, action paths, eye dilation. Eye fixations provide important information to judge the relevancy level of any document during a web search. Fixation refers to a continuous stable gaze for approximately 150-320 milliseconds. Pupil dilation is a measure utilized to indicate people's arousal or interest in the web page [9].

From the literature survey, we observed that over the past two decades researchers have shifted their interest from the manual approach of page quality-related judgments collection to an automatic approach to page quality-related judgments collection. We tried to find out the reasons for this change. We found that there are several advantages of the automatic approach over the manual approach used for collecting page quality-related judgments. These advantages are summarized in Table 11.2.

Table 11.2: Statistical approaches vs. automatic approaches.

Statistical Methods	Automatic Methods
Statistical methods require relevance judgments from experts and searchers to prioritize the web pages in search engines database.	Automatic methods use the users' interaction with browsers to assess the quality of web documents.
Statistical methods require extra cost for expensive expert's judgments.	Automatic methods require low cost for evaluation of search engines.
Not possible to collect real time data.	It is possible to collect real time data.
It requires more time period to evaluate the quality of web pages.	It does not require additional time period to evaluate the quality of web pages.
Small group of people requires to take decision to evaluate and assigned numeric score on each web page.	All searchers require to take decision to evaluate and assigned numeric score on each web page.
Searchers role becomes partial in relevance scores computation.	Searchers have big role to decide the usefulness of web-documents.
Experts know their contribution in evaluation of web-documents.	Searchers always do not knows about the hidden judgments collection.

11.3.8 Coverage

A statistical sampling method for measuring overlap among search engines and their relative coverage has been developed. The two metrics proposed here are used to measure the relative ratio of coverage and overlap only for the results of a given query. The metrics are: number of unique hits, which measures overlap, and the relative number of returning hits to total hits in a given domain, which measures a relative ratio of coverage for a given query. These measures are comparable, in that they are applicable to comparisons of web search engines.

11.3.9 Response Time

This is the time period that starts from the point of query submission till user gets a huge list of response results. Response time is directly related to the activeness of search engines and kept to a minimum as much as possible.

11.4 Factors Affecting Search Engines

11.4.1 Evaluation

How information retrieval systems carry out their search and select the relevant documents corresponding to the submitted queries completely depends upon the system's underlying design philosophy. Searching the inverted index for web pages, meeting the query requirements referred to as "matching," is typically a graded relevance search. Having determined which subset of web pages meets the query requirements to some degree, normally a similarity score is computed for the queries and documents. This similarity score is computed with the help of an underlying algorithm. After computing the similarity of each web page in the set of relevant documents, the information retrieval system presents an ordered list in which the documents are ranked according to the decreasing level of their relevance scores. The sophistication of this ordered list again depends on the algorithm being used. So, the underlying algorithm becomes a very important part of any search engine.

We mainly focus on four factors [5] that directly and/or indirectly affect the information retrieval evaluation process. These factors are 1) query formulation, 2) user feedback to web page, 3) w3 rules, and 4) web developers' fake techniques.

11.4.2 Query Formulation

The utilization of information retrieval systems that help searchers locate the relevant information on the web are increasing with time. Such systems attract users with a need for information or to enhance their current state of knowledge. Typically, searchers make attempts to express this need with a set of query terms submitted to the search systems. These query terms are compared to each object in the collection and a set of close matching objects are retrieved. These objects may be completely relevant, partially relevant or somewhat relevant. It is also possible that the system may retrieve objects without having any relevant information at all.

Query reformulation [10] is an essential part of successful information retrieval. A "query" is a collection of one or more searching keywords taken from the natural language that include logical operators and modifiers. A "keyword" is basically a string of characters without any space. It is quite amazing that very few searchers seldom use logical operators and modifiers in their queries. In a study conducted in two search engines named WebCrawler and Magellan, the author found that only 12% of the queries out of 2000 queries contain Boolean operators. Hoechstoetter and Koch presented an analysis with the search engine Fireball [11]. The authors found that only three percent out of sixteen million queries contained Boolean operators and only eight percent contained a phrase operator. Silverstein *et al.* [12] verified that only a few percent of the queries included advance operators out of the total submitted queries. The information searching process has four categories: problem analysis, query formulation, processing of query and evaluation of result. Searchers are required to identify the problem and arrange the query so that it meaningfully conveys the actual problem logically existing in their minds to the search engines. We should take proper care during query formulation because a tiny mistake changes the direction of the searching process. Most of the new users make mistakes during query formulation; as a result they fail to obtain better results. Searchers must be aware of the fact that the query does not uniquely identify a single web page in the collection. Instead, several web documents may match the query with their own degree of relevance.

11.4.3 User Feedback to a Web Page

Information requirements can be dynamic and may change dramatically in gradual ways in a search session. The rapid growth in web search traffic makes search activity logs a more valuable source of information for understanding user perceptions which can be used further for many practical tasks, including re-formation of rankings. Implicit user's feedback enables a search engine to judge the relevance level of available web pages on the web efficiently at zero cost. This technique, which collects real-time data and judgments, holds the promise of doing retrieval evaluation faster. This technique utilizes the user's behavior within a single web page, such as save, copy, print, add to favorites, hyperlinks clicked, the amount of scrolling and mouse activity, to judge the relevance level. In explicit user feedback method, experts' judgments are required to fix the relevance status to web pages, which is a challenging task because it is too expensive to hire experts to collect judgments for the search engine evaluation process and it is not guaranteed that the normal searchers will agree with these judgments. Another major problem with experts' judgments is slow evaluation. So, it is not practically possible to use the experts' judgments for the evaluation of web pages in such a large and dynamic network "Web." Explicit users' judgments are the creation of a single person from a small group of people. Experts' judge the information, related to their area of knowledge or region. In implicit users' feedback of evaluation, the decisions from each searcher are considered to assign the final ranking or position to web documents. This is not so in explicit feedback method of evaluation.

11.4.4 W3 Rules

There are billions of online websites running on the web for various organizations and businesses across the world. It is the most popular, efficient, fast and cheapest medium to promote a business. To compete with competitors, it is very essential to extend the traffic potential of your website to be a leading online marketer. A web developer is an important person that becomes partially responsible to hike the ranking of any website. Web developers apply w3 rules for web-page designing. These w3 rules also play a major role in marking a search engine as an effective information retrieval system because all search engines take w3 rules of designing into consideration during ranking/indexing the documents.

The various factors that play an important role in searching to compute the relevance scores for web documents are listed below:

- Content should be very powerful
- Proper use of CSS
- HTML coding validation
- Proper use of hyperlinks
- Proper use of meta tags (title, keywords, alt, description)
- Proper use of site map or RSS feed
- The selection of a domain name and its duration
- Avoid broken links

On the medium of the internet, all the web pages are essentially converted into HTML. The search engines visit and evaluate the web pages repeatedly, and assign them top positions which are designed with the w3 guidelines. When the HTML source code of any web page follows all the w3 guidelines, it is known as strong HTML coding.

Few search engines are very popular in the search market, which satisfy the searchers in very short span of time with the desired information. Anyone can raise the question of why these search engines select the web pages which are not following the w3 guidelines and present these pages to the searchers. The answer to this question is that some web pages which are not built through the w3 rules also contain useful information, so these pages cannot be ignored.

11.4.5 Web Developers' Fake Techniques

The Web is a single medium that can advertise your business across the world in a permanent way. But in order to use the advantages of the web in business promotion it is required that the concerned documents should get a position in the top ten citations in the landing page because most of the searchers prefer to visit only a few top citations. Often, optimization of web documents is done to hike the rank of websites. Search engine optimization (SEO) is the process of improving the visibility of a website or a web page for search engines. Optimizing a website may include editing its content, HTML code and associated coding to both to increase its relevance to specific keywords and to remove barriers to the indexing activities of search engines. To optimize a website and hike the rank of web pages, web developers use various fake tactics [12-15]. These tactics are listed below:

- Use of fake keywords while page content does not match with these keywords at all.
- Huge advertisement of any particular website to www so that a large group of people can hit the citation, in order to popularize a particular URL.
- Use of fake title.
- Use of PPC (pay-per-click).
- Words on black background.
- Frequent submission of a particular URL.

Most of the text evaluation techniques are no longer adequate. It is observed that a large number of practices, such as keywords spamming, are opted for by web developers to promote the ranking of web sites. In this way, the relevant pages without optimization are left behind as irrelevant web pages occupying top positions when these are optimized with the fake techniques. A good search engine should be able to identify these types of websites and penalize them by ranking them at the bottom.

Our idea is to develop a search engine evaluation algorithm that can judge the web documents without considering their HTML codes. In other words, we are interested in developing an algorithm in which the impact of HTML source code is not considered during the evaluation of page quality. In this way, search engines will incorporate all the web pages in their evaluation process, whether they are created with w3 guidelines or built normally.

11.5 Conclusion

This research work explored the various methodologies and statistical measures that have been used for better evaluation of search engines. It highlighted the various forums that have been used and provided various data for better evaluation of search engines. The automatic approaches, Click-through data and Eye tracking, were presented along with the reasons as to why the automatic approach for page quality-related judgments collection is considered superior to traditional approaches implemented in this research. In such a way, search engines will incorporate all the web pages in their evaluation process, whether they are created with w3 guidelines or built normally.

References

1. Chu, H., & Rosenthal, M. (1996, October). Search engines for the World Wide Web: A comparative study and evaluation methodology. In *Proceedings of the Annual Meeting-American Society for Information Science* (Vol. 33, pp. 127-135).
2. Singh, J. N., & Dwivedi, S. K. (2014, October). Comparative study on evaluative measures of search engines. In *Proceedings of 3rd International Conference on Reliability, Infocom Technologies and Optimization* (pp. 1-6). IEEE.
3. Bashir, M. (2014). *Optimally selecting and combining assessment and assessor types for information retrieval evaluation* (Doctoral dissertation, Northeastern University Boston).
4. Baeza-Yates, R., & Ribeiro-Neto, B. (1999). Modern information retrieval (Vol. 463). *New York: ACM press.*
5. J. N. Singh and S. K. Dwivedi, Analyze. *Evaluate the Performance of Search Engines.* Lambert Academic Publishing, 2015.
6. Goutam, R. K., & Dwivedi, S. K. (2011, September). Search Engines Evaluation using users efforts. In *2011 2nd International Conference on Computer and Communication Technology (ICCCT-2011)* (pp. 589-594). IEEE.
7. Al-Maskari, A., Sanderson, M., & Clough, P. (2007, July). The relationship between IR effectiveness measures and user satisfaction. In *Proceedings of the 30th Annual International ACM SIGIR Conference on Research and Development in Information Retrieval* (pp. 773-774).
8. Järvelin, K., & Kekäläinen, J. (2002). Cumulated gain-based evaluation of IR techniques. *ACM Transactions on Information Systems (TOIS)*, 20(4), 422-446.
9. Shang, Y., & Li, L. (2002). Precision evaluation of search engines. *World Wide Web*, 5(2), 159-173.
10. Aula, A. (2003, November). Query Formulation in Web Information Search. In *ICWI* (pp. 403-410).
11. Hoechstoetter, N., & Koch, M. (2007). Standard comparators for Information Searching Behaviour in Search Engines. *JIS*, 523(v5).
12. Dogra, V., Verma, S., Jhanjhi, N. Z., Ghosh, U., & Le, D. N. (2022). A Comparative Analysis of Machine Learning Models for Banking News Extraction by Multiclass Classification With Imbalanced Datasets of Financial News: Challenges and Solutions. *International Journal of Interactive Multimedia & Artificial Intelligence*, 7(3).
13. Nguyen, L. D., Le, D. N., & Vinh, L. T. (2014, December). Detecting phishing web pages based on DOM-tree structure and graph matching algorithm. In *Proceedings of the fifth symposium on information and communication technology* (pp. 280-285).
14. Miller, M. P., & Blatnik, J. A. (2022). Evaluation of information on the Internet regarding surgical mesh for hernia repair: analysis of websites found through three popular search engines. *Hernia*, 26(2), 581-587.
15. Silverstein, C., Marais, H., Henzinger, M., & Moricz, M. (1999, September). Analysis of a very large web search engine query log. In *ACM Sigir Forum*, Vol. 33, No. 1, pp. 6-12). New York, NY, USA: ACM.

12

Synthesis and Analysis of Digital IIR Filters for Denoising ECG Signal on FPGA

Seema Nayak[1], Manoj Nayak[2], Shamla Matri[3], Kanta Prasad Sharma[4]

[1] Department of Electronics and Communication Engineering, IIMT College of Engineering, Greater Noida, India
[2] Manav Rachna International Institute of Research and Studies Faridabad, India
[3] School of Computer engineering and Technology, MIT-WPU, Pune, India
[4]Computer Science and Engineering, GLA University Mathura, India
Email: seema_jessica@rediff.com

Abstract

In recent years, field-programmable gate arrays have progressively become vital for creating the means for every type of digital system design due to their compact growth point and lower expense. Moreover, their flexibility has enabled growth in the field and made hardware compatible with runtime environment. This chapter focuses on a new, simple and well-organized approach for synthesis design of optimal order IIR digital filters to reduce noise in ECG signal on FPGA. A summary of its resource consumption (amount of slice, slice flip-flops figure, number of 4-input LUTs, figure of bonded IOBs, number of BUFG and DSP48A1 slice), the timing (smallest amount of input arrival time sooner than clock [set-up time] and maximum output requisite period subsequent to clock [hold time] and power consumed are presented after synthesis and simulation. This is achieved by conversion of MatLab code of different designed IIR digital filters for denoising ECG signal into Verilog code using HDL command line interface. Spartan-6 FPGA (XC6SLX75T with 3FGG676 package) was used as a target device. To estimate power consumed by digital design, Xilinx Power Estimator tool is used. Furthermore, the complexity of a filter structure is also determined on FPGA platform for its suitability as a hardware efficient filter design. The results suggest that the FPGA-based digital filter design reduces the complexity and cost by reducing the number of multipliers and adders, which occupy a small portion of the chip area and consume lower power than in MatLab, hence are suitable for ECG portable devices.

Keywords: FPGA, ECG signal, MatLab

12.1 Introduction

Field-programmable gate arrays (FPGA) are prefabricated silicon chips that can be programmed electrically in order to design various types of digital systems [1]. They are a low-cost solution with fast response to the market compared to application-specific integrated circuit (ASIC), which generally uses a lot of resources, i.e., time and money, to obtain a device. For altering needs, a sector of FPGA can be partly reconfigured while the rest of an FPGA is still in a row. The design of FPGAs alters from seller to seller and is characterized by arrangement, logic block content and routing resources. The designers are inclined more towards FPGAs as modern architectures, with memory blocks, embedded processors and digital signal processors (DSPs) being developed in FPGA.

IIR filters have been implemented on FPGA, and the structure has added compensation, such as complete adaption in FPGA arrangement, which leads to the high throughput, filtering algorithm, effectiveness of hardware utilization, achieving high accuracy rate. The digital filters require many adders, multipliers and registers. Logical array can be used to form the adders and multipliers. FPGA is a structured internal logic array with loaded link resources, and it is more appropriate for hardware realization of digital filter. The issues and fundamental challenges occur in programmable routing of circuit design and architecture. Digital filter implementation on FPGA offers a superior performance by reducing the complexity of filter structure, hardware requirements and enhancing speeds. An application of general purpose multipliers is a conventional move. The performance of implemented multipliers on FPGA architecture does not allow constructing high-performance digital filters. Parallelism nature of FPGA provides improvement in speed, less resource usage and low power consumption [3]. A large number of multipliers are required in the implementation of digital filters in the hardware of FPGA which add the complexity in the circuit. In digital filters the reduction of multipliers is the key to minimizing the complexity of hardware. The goal is to implement IIR on FPGA by creating Verilog code, which introduces innovation in design synthesis and analysis by HDL coders. Here the optimal order IIR (Chebyshev I & II, Butterworth and Elliptic) filters are selected based on mean square error (MSE) and signal to noise ratio (SNR) for denoising ECG signal in MATLAB initially before implementation on FPGA [3].

The main challenge, however, is in using digital filters on hardware to achieve rapid speed at low hardware costs. Therefore, it is imperative to choose operating methods and tools carefully based on design specifications in order to save a lot of time and effort. The use of digital filters on an FPGA shows how adaptable the method is and how it performs better than more traditional methods, which can save a lot of time and work. The digital filters implementation on FPGA illustrates that the approach is flexible and comparably or superior to the conventional approach [4].

12.2 Literature Survey

With the advancement in very large-scale integration (VLSI), realization of digital filters is done in ASIC circuits and FPGA platform. Researchers are actively working towards an approach to design and execute digital filter algorithms made on FPGAs due to their inherent advantages in their design technology.

Chou *et al.* [4] provided a way to implement algorithms on digital filter based on FPGAs. Kuon *et al.* [5] reviewed and surveyed the development of programmable logic devices and presented the fundamental programming techniques upon which programmability is based, followed by the definition of the considerations learned from the study of architecture. Ravikumar [6] carried out digital filtering with low pass FIR to filter the 50 Hz coupled noise and other noises of high frequency. A Xilinx ChipScope was the tool used to test the results and FPGA was used to run the logic with FPGA development board, Xilinx Spartan-3 family. Islam *et al.* [7] proposed the architecture of a programmable digital IIR filter-based Xilinx FPGA board. Dixit and Gupta [8] proposed the implementation and simulation of IIR filter using Xilinx System Generator software and the Simulink environment in MATLAB on an FPGA. They suggested that the capability of the FPGA greatly increases due to parallel processing and the speed of performance in the functioning of the digital filter.

Kasetwar and Gulhane [9] offered a study on performance of adaptive power line interference canceler for ECG signals and least mean-square (LMS) algorithm recommended for implementation of adaptive power line interference canceler. Pawar and Bhaskar [10] used high-pass filtering of ECG signal and then implemented it on FPGA platform. Yadav and Mehra [11] examined the cost performance of different IIR filters on the basis of implementation which is designed for real-time application. The cost of implementation was arrived at based on the filter order, multiplier, adder, and input samples. The implementation and synthesis of digital IIR filters on FPGA offer hardware utilization effectiveness, high throughput and high rate of precise calculation [8, 12].

A study on ECG signal processing with FIR filter architecture using VHDL was researched by Ravikumar [6] and Narsale *et al.* [12], and the same study using Verilog and Xilinx was done by Jairath *et al.* [13], Vijaya *et al.* [14], Singh *et al.* [15], and Gaikwad [16] to generate an HDL command for implementation of an inverse sinc filter on FPGA. Kumar and Meduri [17] generated HDL code with distributed arithmetic architecture of high order matched filter by using the same technique. A ModelSim 6.4a simulator was used for the simulation of generated test benches of the optimal order filters. Studies in [6, 12, 14] reported simulation of FPGA-based design using ModelSim software. The present work used ModelSim 6.4a to generate test benches for digital filters with optimal order for denoising ECG signal. A series of Xilinx FPGA families such as Spartan-3, Vertex-4, and Spartan-6, etc., are commercially available for synthesis of generated HDL code of digital filter on FPGA.

Moschetta [18] proposed a novel configuration combining rotors and fixed winding to gain efficiency and maneuverability at low speeds. Meftah *et al.* [19] modeled a surface micro-mechanical accelerometer from a silicon MEMS accelerometer. Singh and Bansal [20] used acoustic-based inspection methods for on-field screening of containers, which hold high importance in the fields of security and defense. Zhang and Yao [21] proposed a fuzzy Petri nets method for modeling and analyzing the reliability of a multi-state software system. Wang *et al.* [22] presented a novel fetal ECG blind source extraction algorithm based on blind source separation in noise. The algorithm used was discrete wavelet transformation and moved the conventional time-domain signals to the wavelet-domain to reduce the noise in FECG. Bhogeshwar *et al.* [23] used digital filter methods to deal with the noise artefact in ECG signal and the research was carried out from the available MIT-BIH arrhythmia database instead of taking real-time acquired signal from human object as it was informal and a less time-consuming method [24]. As previously discussed, the literature survey has shown that the high-end FPGA has a huge throughput advantage but comes with hardware costs.

12.3 Methods and Materials

This section discusses the methodology for execution of the FPGA-based optimal order IIR digital filters for denoising ECG signal. The workflow chart of the methodology is shown in Figure 12.1.

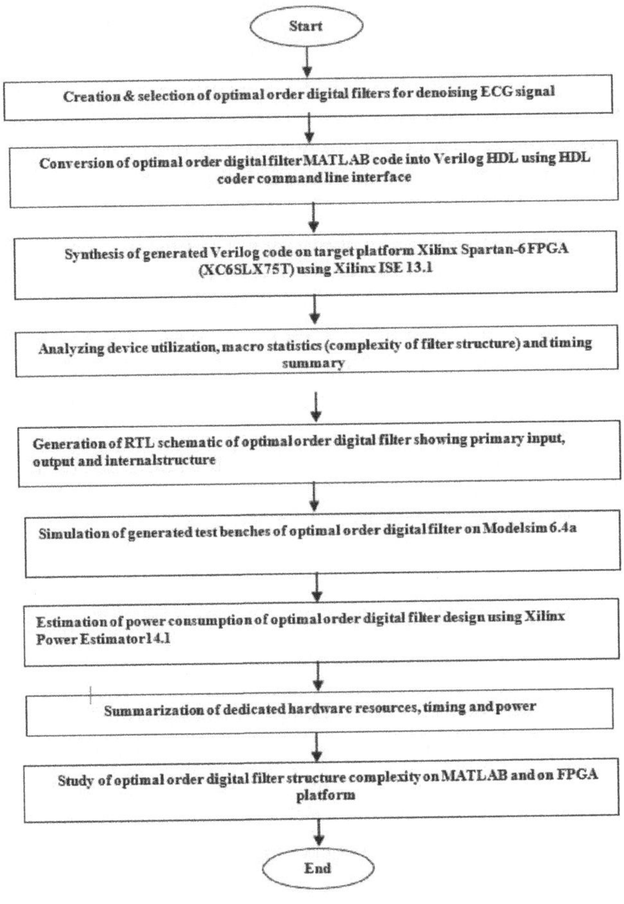

Figure 12.1: Workflow chart for synthesis of digital filters on FPGA.

12.3.1 Conversion of MATLAB Code into Verilog HDL

The first IIR digital filters were designed in MATLAB for denoising ECG signal and selected optimal order filter based on signal SNR and MSE [3, 25-31]. In addition, their structural complexity based on multipliers and adder in MATLAB was studied.

MATLAB command line is used to generate filter design HDL code and test bench for a quantized filter using its property name and its value. All the properties have their default values, which can be changed to customize the output. Filter Design HDL Coder automatically creates Verilog test benches for quick simulation, testing, and verification of the generated HDL code. Thus, synthesis of the design is taken out in Xilinx ISE 13.1 with Xilinx Spartan-6 (XC6SLX75T) target device and the generated test benches are simulated in ModelSim 6.4a software [32-35].

12.3.2 Synthesis of Digital Filters on FPGA

The performance of digital filters on FPGA illustrates that the approach is flexible and superior. After designing and selecting the optimal order digital filters based on MATLAB, Xilinx 13.1 package with target device Spartan-6 (XC6SLX 75T) [33] is used for synthesis of generated Verilog code of optimal order filter on FPGA. The workflow chart of the synthesis process of Verilog code of optimal order IIR and FIR filters is shown in Figure 12.1.

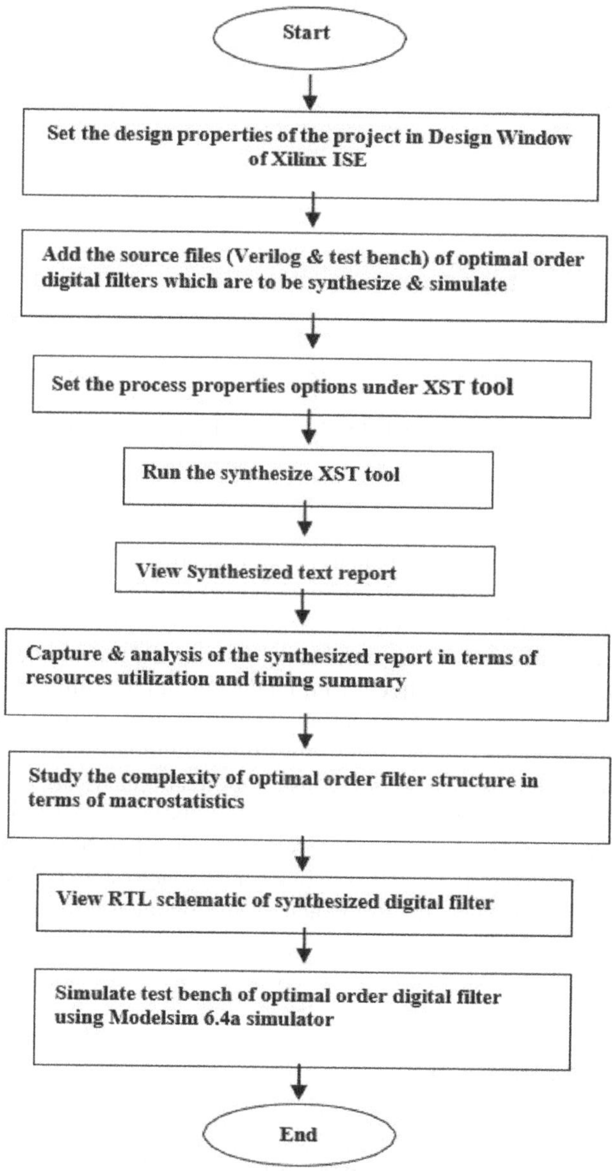

Figure 12.2: Workflow chart of design steps for synthesis and simulation of digital filters.

The design properties of the project in the Design Window of Xilinx ISE are shown in the snapshot Figure 12.3 [32].

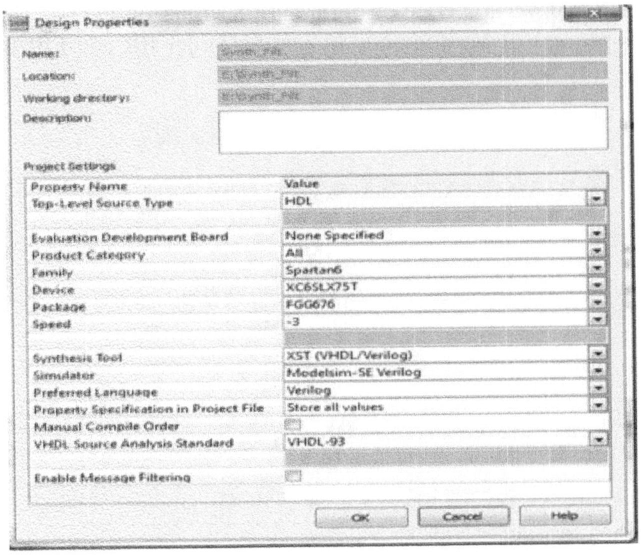

Figure 12.3: Snapshot of design property window.

The source files (Verilog and test bench) of optimal order digital filters are added, which are to be synthesized and simulated followed by setting the Process Properties options under XST synthesis tool. After synthesis, the resource utilization summary in number of slice flip flops, number of slices, number of 4-input LUTs, number of bonded IOBs, number of BUFG and DSP48A1 slice for initial design and the timing summary in respect to smallest amount input advent time earlier than clock (set-up time) and utmost output requisite time subsequent to clock (hold time) for preliminary design, are captured and displayed. From the details of resource consumption summary, the resources to be used for the target FPGA are analyzed and from the reports in respect to delays (set-up and hold time), maximum frequency and timing summary are also studied. The complexity of filter structure is studied on FPGA platform in terms of macro statistics (adders/subtractors, multipliers and registers). Then, the test bench of Verilog file is simulated using ModelSim simulator.

12.4 Results and Discussion

The complexity of hardware for optimal order digital IIR filters structure is very large in conditions of multipliers, adders and delays using MATLAB. A closing outline report on the strengths, weaknesses, opportunities, and threats (SWOT) of difficulty of the structure of optimal order filters is presented in Table 12.1 [3].

Table 12.1: Summary of IIR filters structure information in MATLAB [3].

SN	Type of Filter	Order	Multiplier	Adder	States	Multiplication Per Input sample	Addition Per Input Sample
1	Butterworth	5	10	10	5	10	10
2	Chebyshev-I	3	6	6	3	6	6
3	Chebyshev-II	5	10	10	5	10	10
4	Elliptic	3	6	6	3	6	6

Furthermore, the design of optimal order IIR digital filters is synthesized and analyzed in the present work.

Synthesis reports and simulation results of the traditional approach of designing IIR digital filters are presented. The resource utilization summary (slice LUTs, number of slice registers, bonded IOBs and BUFG/BUFGCTRLS, fully used LUT-FF pairs), macro statistics summary (adders/subtrators, adder tree and registers), final registers and timing summary (minimum period, maximum frequency, setup time, hold time), RTL schematic digital filter, RTL diagram of internal structures, and simulation waveform and power estimation report (Xilinx Power Estimator User Guide UG440 (v13.4) January 18, 2012) of optimal order Butterworth, Chebyshev II, Chebyshev I, and Elliptic digital filter are presented. Furthermore, hardware resources to be used in FPGA, complexity of filter structure, speed and total estimated power of the traditional approach of designed digital filters are tabulated.

12.4.1 Butterworth Filter

The optimal order of Butterworth filter for making noise-free ECG signal is selected on the basis of MSE and SNR and is 5 (see Table 12.1). Synthesis reports of Butterworth filter after synthesis on FPGA are given in Figure 12.4 to Figure 12.9.

```
Device utilization summary:
---------------------------

Selected Device : 6slx75tfgg676-3


Slice Logic Utilization:
 Number of Slice Registers:              109  out of  93296     0%
 Number of Slice LUTs:                   254  out of  46648     0%
    Number used as Logic:                254  out of  46648     0%

Slice Logic Distribution:
 Number of LUT Flip Flop pairs used:     349
   Number with an unused Flip Flop:      240  out of    349    68%
   Number with an unused LUT:             95  out of    349    27%
   Number of fully used LUT-FF pairs:     14  out of    349     4%
   Number of unique control sets:          1

IO Utilization:
 Number of IOs:                           35
 Number of bonded IOBs:                   35  out of    348    10%

Specific Feature Utilization:
 Number of BUFG/BUFGCTRLs:                 1  out of     16     6%
 Number of DSP48A1s:                      11  out of    132     8%
```

Figure 12.4: Device utilization summary for Butterworth filter.

```
Advanced HDL Synthesis Report

Macro Statistics
# MACs                                : 5
 16x14-to-30-bit MAC                  : 1
 16x15-to-30-bit MAC                  : 2
 16x16-to-30-bit MAC                  : 2
# Multipliers                         : 3
 16x15-bit multiplier                 : 2
 16x16-bit multiplier                 : 1
# Adders/Subtractors                  : 14
 30-bit adder                         : 3
 33-bit adder                         : 3
 34-bit adder                         : 8
# Registers                           : 112
 Flip-Flops                           : 112
```

Figure 12.5: Summary of macro statistics for Butterworth filter.

```
Final Register Report

Macro Statistics
# Registers                           : 112
 Flip-Flops                           : 112
```

Figure 12.6: Register report for Butterworth filter.

```
Timing Summary:
---------------
Speed Grade: -3

   Minimum period: 60.454ns (Maximum Frequency: 16.541MHz)
   Minimum input arrival time before clock: 3.552ns
   Maximum output required time after clock: 3.701ns
   Maximum combinational path delay: No path found

Timing Details:
---------------
All values displayed in nanoseconds (ns)
```

Figure 12.7: Timing summary for Butterworth filter.

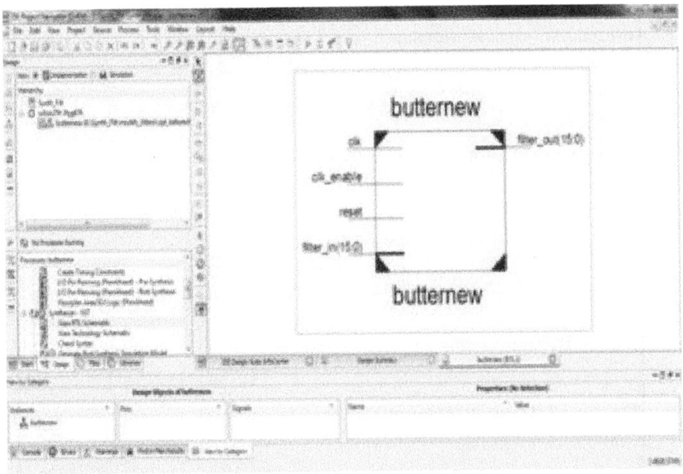

Figure 12.8: RTL schematic for Butterworth filter.

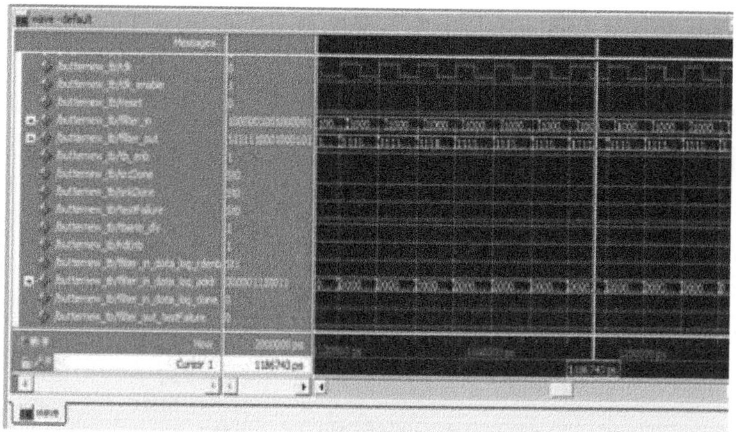

Figure 12.9: RTL internal structure for Butterworth filter.

Test bench simulation result is given in Figure 12.10 and the power estimator report in Figure 12.11.

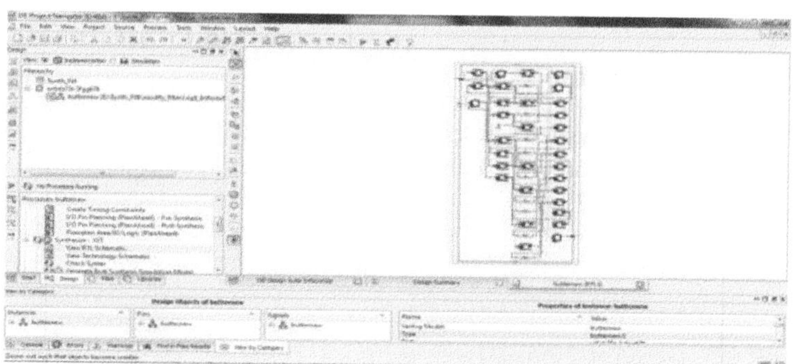

Figure 12.10: ModelSim simulation result for Butterworth filter.

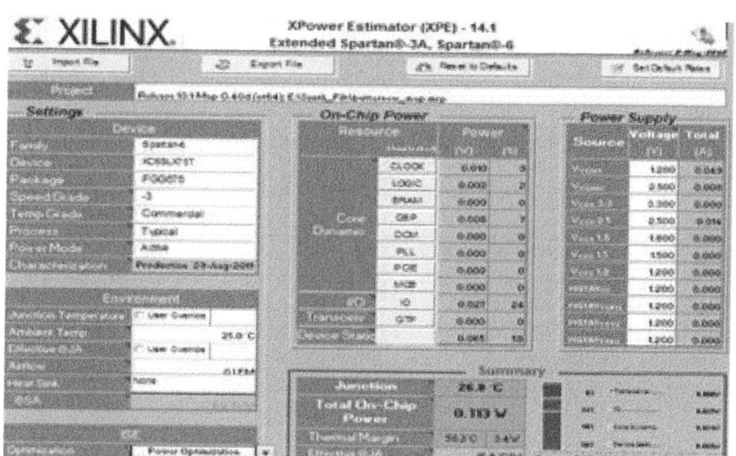

Figure 12.11: Power estimation for Butterworth filter.

After synthesis and simulation, all the reports are tabulated in Table 12.2 [25].

Table 12.2: Summary of resources for Butterworth filter.

SN	Parameters	FPGA Implementation	
1	No. of Slices Register	109 out of 93296	0%
2	No. of Slice LUTs	254 out of 46648	0%
3	No. of fully used LUT-FF pairs	14 out of 349	4%
4	No. of bonded IOBs	35 out of 348	10%
4	No. of BUFG/BUFGCTRLs	1 out of 16	6%
6	No. of DSP48A1S	11 out of 132	8%
7	Minimum Period	60.454 ns	
8	Maximum frequency	16.541 MHz	
9	Minimum input arrival time before clock (Ts)	3.552 ns	
10	Maximum output required time after clock (TH)	3.701 ns	
11	Power	113 mw	
12	Macro Statistics	MACs-5,Multipliers-3, Adder/Subtractor-14, Regs-112	

12.4.2 Chebyshev-I Filter

The optimal order of Chebyshev-I filter for noise-free ECG signal is selected as order 3 (Table 12.1). Synthesis reports of Chebyshev-I filter after synthesis on FPGA is given in Figure 12.12 to Figure 12.17.

```
Device utilization summary:
---------------------------

Selected Device : 6slx75tfgg676-3


Slice Logic Utilization:
 Number of Slice Registers:                 66  out of  93296    0%
 Number of Slice LUTs:                      75  out of  46648    0%
    Number used as Logic:                   75  out of  46648    0%

Slice Logic Distribution:
 Number of LUT Flip Flop pairs used:       129
   Number with an unused Flip Flop:         63  out of    129   48%
   Number with an unused LUT:               54  out of    129   41%
   Number of fully used LUT-FF pairs:       12  out of    129    9%
   Number of unique control sets:            1

IO Utilization:
 Number of IOs:                             35
 Number of bonded IOBs:                     35  out of    348   10%

Specific Feature Utilization:
 Number of BUFG/BUFGCTRLs:                   1  out of     16    6%
 Number of DSP48A1s:                         8  out of    132    6%
```

Figure 12.12: Device utilization summary for Chebyshev-I filter.

```
Advanced HDL Synthesis Report

Macro Statistics
# MACs                                              : 3
 16x15-to-35-bit MAC                                : 2
 16x16-to-35-bit MAC                                : 1
# Multipliers                                       : 1
 16x16-bit multiplier                               : 1
# Adders/Subtractors                                : 10
 21-bit adder                                       : 1
 30-bit adder                                       : 4
 35-bit adder                                       : 4
 37-bit adder                                       : 1
# Registers                                         : 70
 Flip-Flops                                         : 70
```

Figure 12.13: Summary of macro statistics for Chebyshev-I filter.

```
Final Register Report

Macro Statistics
# Registers                                    : 66
 Flip-Flops                                    : 66
```

Figure 12.14: Register report for Chebyshev-I filter.

```
Timing Summary:
---------------
Speed Grade: -3

   Minimum period: 21.038ns (Maximum Frequency: 47.534MHz)
   Minimum input arrival time before clock: 3.305ns
   Maximum output required time after clock: 3.732ns
   Maximum combinational path delay: No path found

Timing Details:
---------------
All values displayed in nanoseconds (ns)
```

Figure 12.15: Timing summary for Chebyshev-I filter.

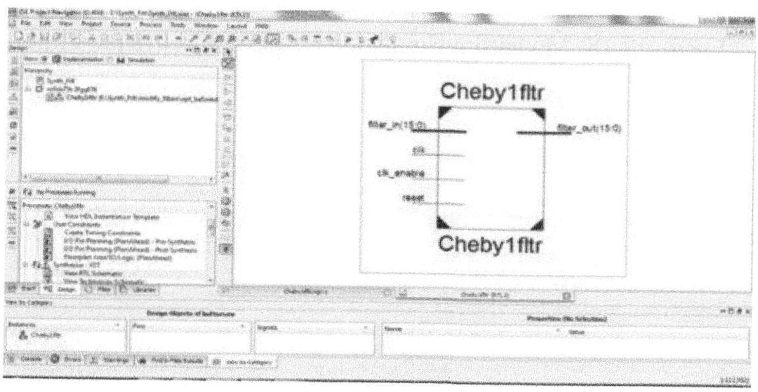

Figure 12.16: RTL schematic for Chebyshev-I filter.

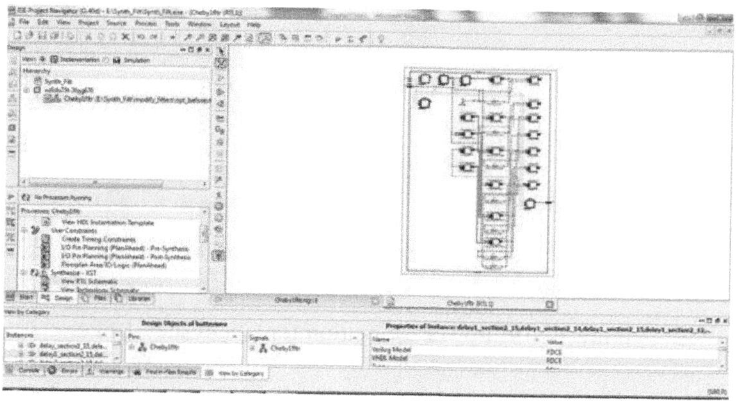

Figure 12.17: RTL internal structure for Chebyshev-I filter.

Test bench simulation result is given in Figure 12.18 with power estimator report in Figure 12.19.

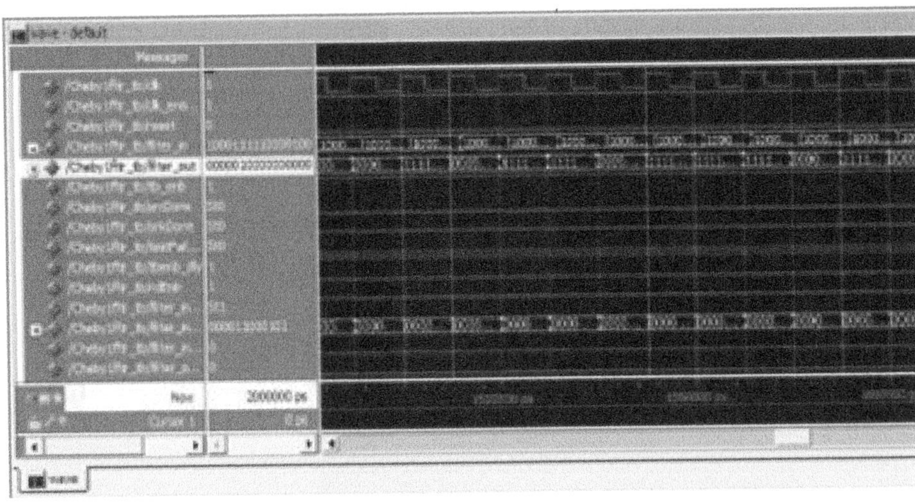

Figure 12.18: ModelSim simulation result for Chebyshev-I filter.

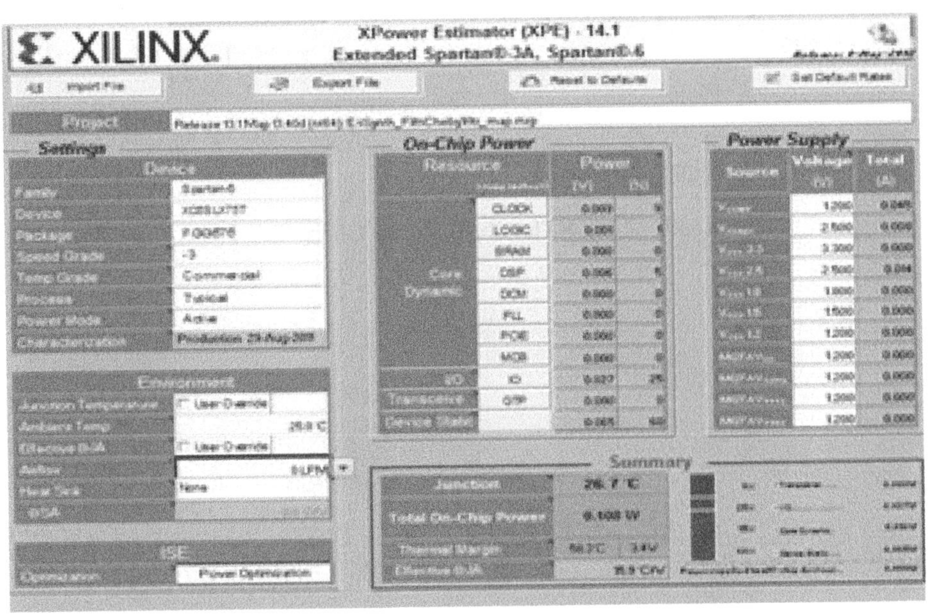

Figure 12.19: Power estimation for Chebyshev-I filter.

After synthesis and simulation, all the reports are tabulated in Table 12.3.

Table 12.3: Summary of resources for Chebyshev-I filter.

SN	Parameters	FPGA Implementation
1	No. of Slices Registers	66 out of 93296 0%
2	No. of Slice LUTs	75 out of 46648 0%
3	No. of fully used LUT-FF pairs	12 out of 129 9 %
4	No. of bonded IOBs	35 out of 348 10%
4	No. of BUFG/BUFGCTRLs	1 out of 16 6%
6	No. of DSP48A1S	8 out of 132 6%
7	Minimum Period	21.038 ns
8	Maximum frequency	47.534 MHz
9	Minimum input arrival time before clock (Ts)	3.305 ns
10	Maximum output required time after clock (TH)	3.732 ns
11	Power	108 mw
12	Macro Statistics	MACs-3,Multiplier-1, Adder/Subtractor- 10,Regs-66

12.4.3 Chebyshev-II Filter

The optimal order of Chebyshev-II filter for denoising ECG signal is selected as order 5 (Table 12.1). Synthesis and simulation reports of Chebyshev-II filter after synthesis on FPGA are given in Figure 12.20 to Figure 12.26 with power estimator report in Figure 12.27.

```
Device utilization summary:
---------------------------

Selected Device : 6slx75tfgg676-3

Slice Logic Utilization:
 Number of Slice Registers:              97  out of  93296     0%
 Number of Slice LUTs:                   91  out of  46648     0%
    Number used as Logic:                91  out of  46648     0%

Slice Logic Distribution:
 Number of LUT Flip Flop pairs used:    175
   Number with an unused Flip Flop:      78  out of    175    44%
   Number with an unused LUT:            84  out of    175    48%
   Number of fully used LUT-FF pairs:    13  out of    175     7%
   Number of unique control sets:         1

IO Utilization:
 Number of IOs:                          35
 Number of bonded IOBs:                  35  out of    348    10%

Specific Feature Utilization:
 Number of BUFG/BUFGCTRLs:                1  out of     16     6%
 Number of DSP48A1s:                     13  out of    132     9%
```

Figure 12.20: Device utilization summary for Chebyshev-II filter.

```
Advanced HDL Synthesis Report

Macro Statistics
# MACs                                    : 7
 16x12-to-34-bit MAC                      : 1
 16x14-to-34-bit MAC                      : 2
 16x15-to-34-bit MAC                      : 3
 16x16-to-34-bit MAC                      : 1
# Multipliers                             : 1
 16x16-bit multiplier                     : 1
# Adders/Subtractors                      : 13
 20-bit adder                             : 1
 30-bit adder                             : 6
 34-bit adder                             : 5
 37-bit adder                             : 1
# Registers                               : 100
 Flip-Flops                               : 100
```

Figure 12.21: Summary of macro statistics for Chebyshev-II filter.

```
Final Register Report

Macro Statistics
# Registers                                                    : 97
 Flip-Flops                                                    : 97
```

Figure 12.22: Register report for Chebyshev-II filter.

```
Timing Summary:
---------------
Speed Grade: -3

   Minimum period: 23.704ns (Maximum Frequency: 42.187MHz)
   Minimum input arrival time before clock: 3.509ns
   Maximum output required time after clock: 3.701ns
   Maximum combinational path delay: No path found

Timing Details:
---------------
All values displayed in nanoseconds (ns)
```

Figure 12.23: Timing summary for Chebyshev-II filter.

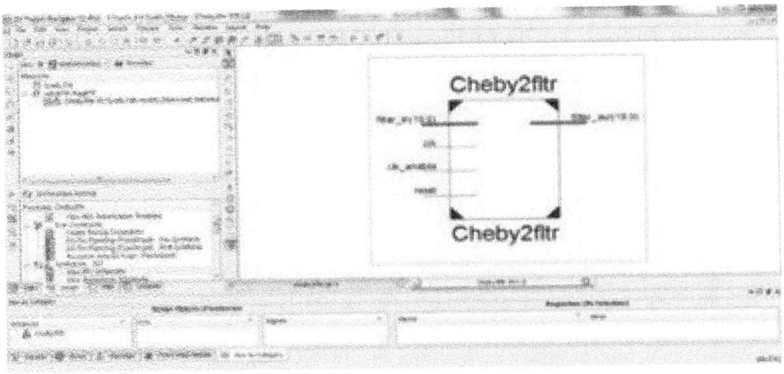

Figure 12.24: RTL schematic for Chebyshev-II filter.

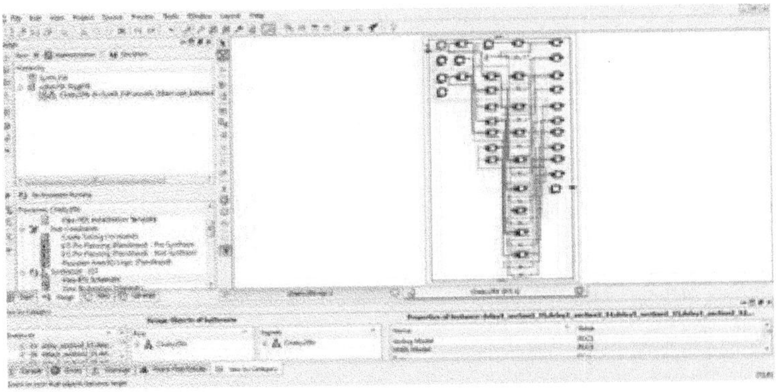

Figure 12.25: RTL internal structure for Chebyshev-II filter.

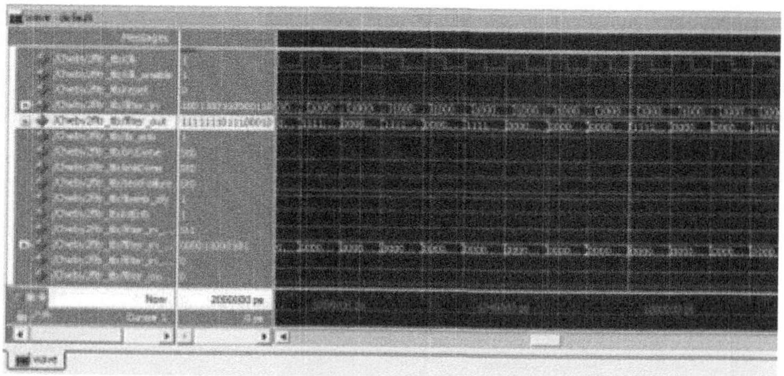

Figure 12.26: ModelSim simulation result for Chebyshev-II filter.

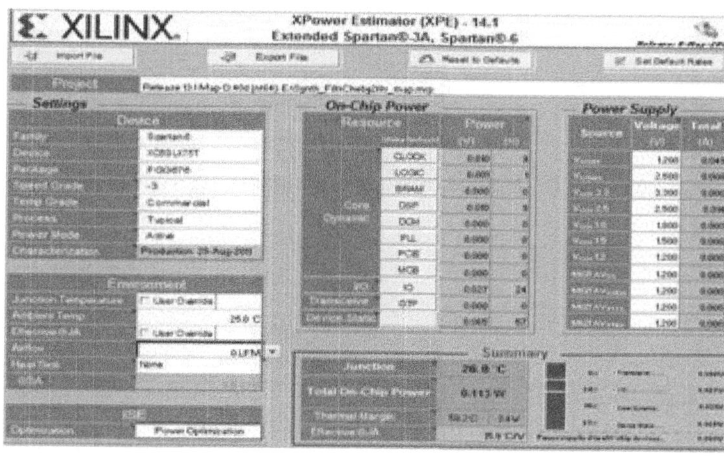

Figure 12.27: Power estimation for Chebyshev-II filter.

After synthesis and simulation, all the reports are tabulated in Table 12.4.

Table 12.4: Summary of resources for Chebyshev-II filter.

SN	Parameters	FPGA Implementation
1	No. of Slices Registers	97 outof93296 0%
2	No. of Slice LUTs	91 outof46648 0%
3	No. of fully used LUT-FF pairs	13 out of 175 7 %
4	No. of bonded IOBs	35 outof348 10%
4	No. of BUFG/BUFGCTRLs	1 outof16 6%
6	No. of DSP48A1S	13 outof132 9%
7	Minimum Period	23.704ns
8	Maximum frequency	42.187MHz
9	Minimum input arrival time before clock (Ts)	3.509 ns
10	Maximum output required time after clock (TH)	3.701 ns
11	Power	113 mw
12	Macro Statistics	MACs-7, Multiplier-1, Adder/Subtractor- 13, Regs-97

12.4.4 Elliptic Filter

The Elliptic filter with optimal order of 3 for denoising ECG signal is selected (Table 12.1). Synthesis and simulation reports of Elliptic filter after synthesis on FPGA is given in Figure 12.28 to Figure 12.34 with power estimator report in Figure 12.35.

```
Device utilization summary:
---------------------------

Selected Device : 6slx75tfgg676-3

Slice Logic Utilization:
 Number of Slice Registers:                 78  out of  93296     0%
 Number of Slice LUTs:                      95  out of  46648     0%
    Number used as Logic:                   95  out of  46648     0%

Slice Logic Distribution:
 Number of LUT Flip Flop pairs used:       159
   Number with an unused Flip Flop:         81  out of    159    50%
   Number with an unused LUT:               64  out of    159    40%
   Number of fully used LUT-FF pairs:       14  out of    159     8%
   Number of unique control sets:            1

IO Utilization:
 Number of IOs:                             35
 Number of bonded IOBs:                     35  out of    348    10%

Specific Feature Utilization:
 Number of BUFG/BUFGCTRLs:                   1  out of     16     6%
 Number of DSP48A1s:                        10  out of    132     7%
```

Figure 12.28: Device utilization summary for Elliptic filter.

```
Advanced HDL Synthesis Report

Macro Statistics
# MACs                                              : 4
 16x15-to-30-bit MAC                                : 2
 16x16-to-30-bit MAC                                : 1
 16x16-to-34-bit MAC                                : 1
# Multipliers                                       : 2
 16x13-bit multiplier                               : 1
 16x16-bit multiplier                               : 1
# Adders/Subtractors                                : 8
 30-bit adder                                       : 2
 32-bit adder                                       : 2
 34-bit adder                                       : 4
# Registers                                         : 80
 Flip-Flops                                         : 80
```

Figure 12.29: Summary of macro statistics for Elliptic filter.

```
Final Register Report

Macro Statistics
# Registers                                         : 78
 Flip-Flops                                         : 78
```

Figure 12.30: Register report for Elliptic filter.

```
Timing Summary:
---------------
Speed Grade: -3

   Minimum period: 46.263ns (Maximum Frequency: 21.616MHz)
   Minimum input arrival time before clock: 3.384ns
   Maximum output required time after clock: 3.668ns
   Maximum combinational path delay: No path found

Timing Details:
---------------
All values displayed in nanoseconds (ns)
```

Figure 12.31: Timing summary for Elliptic filter.

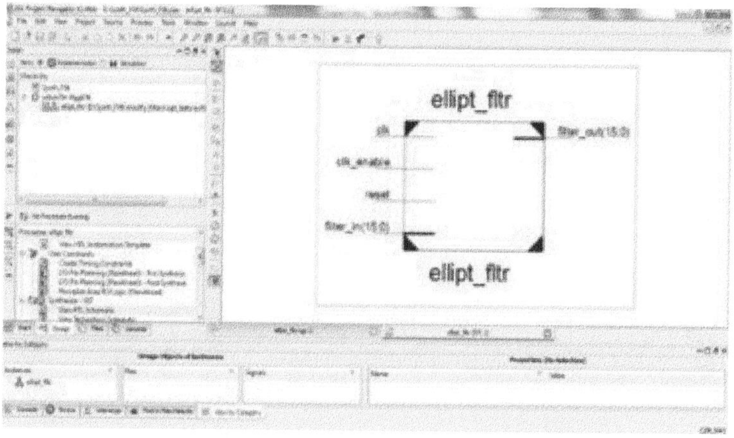

Figure 12.32: RTL schematic for Elliptic filter.

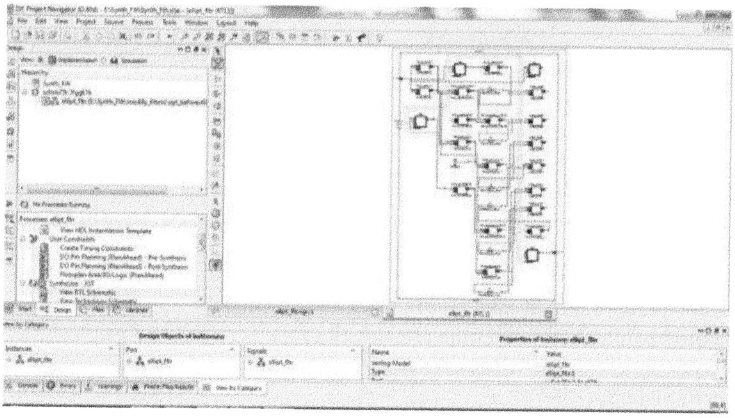

Figure 12.33: RTL internal structure for Elliptic filter.

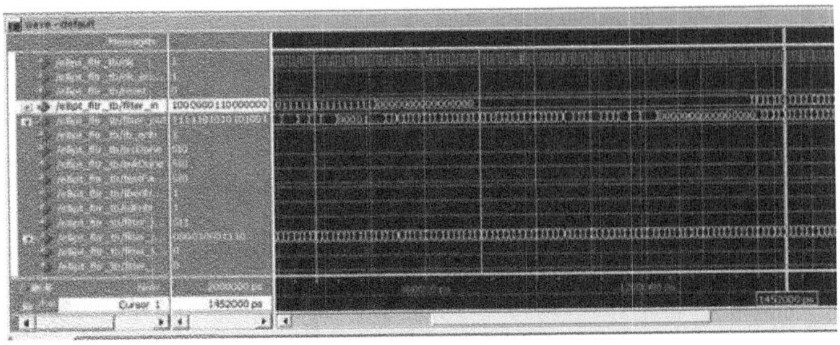

Figure 12.34: ModelSim simulation result for Elliptic filter.

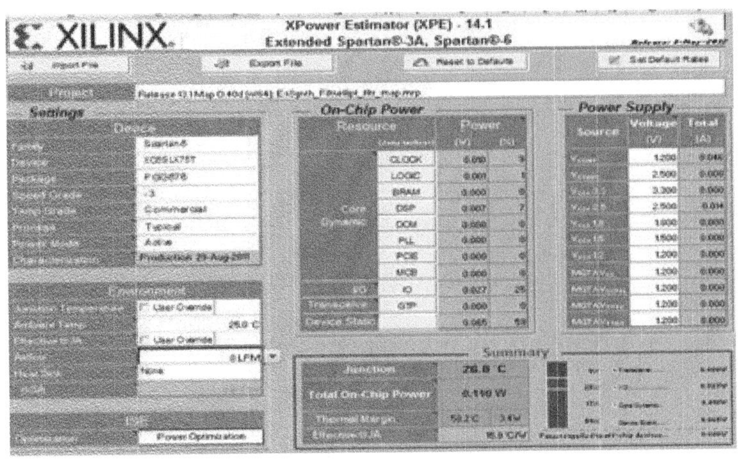

Figure 12.35: Power estimation for Elliptic filter.

All the reports are tabulated in Table 12.5.

Table 12.5: Summary of resources for Elliptic filter.

SN	Parameters	FPGA Implementation
1	No. of Slices Registers	78 out of 93296 0%
2	No. of Slice LUTs	95 out of 46648 0%
3	No. of fully used LUT-FF pairs	14 out of 159 8%
4	No. of bonded IOBs	35 out of 348 10%
5	No. of BUFG/BUFGCTRLs	1 out of 16 6%
6	No. of DSP48A1S	10 out of 132 7%
7	Minimum Period	46.263ns
8	Maximum frequency	21.616 MHz
9	Minimum input arrival time before clock (Ts)	3.384 ns
10	Maximum output required time after clock (TH)	3.668 ns
11	Power	110 mw
12	Macro Statistics	MACs-4, Multiplier-2, Adder/Subtractor- 8, Regs-78

The comparison of basic elements for IIR digital filters is done in FPGA implementation with MATLAB and is given in Table 12.5. After synthesis of digital IIR filters on FPGA platform, it is observed from Table 12.5 that the number of multipliers has been reduced with increase in adder/subtractors and registers, which is shown graphically in Figure 12.36.

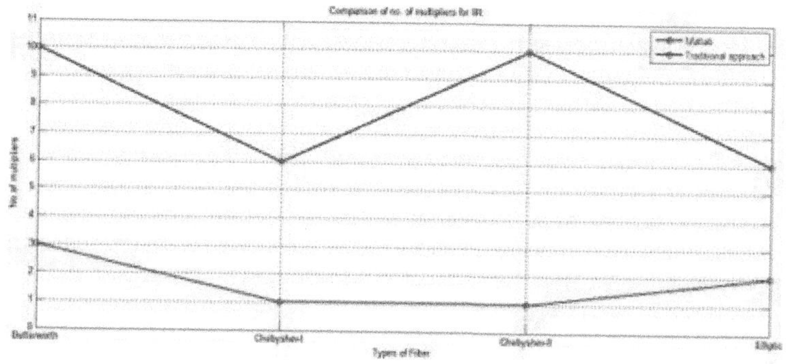

Figure 12.36: Number of multipliers in IIR digital filters (MATLAB and FPGA implementation).

12.5 Conclusion and Future Scope

There are several issues of concern related to area, speed and power for FPGA-based design. The speed, area, and power with high order filters are affected because of complex computation. The complexity of structure for various optimal order IIR digital filters is realized using MATLAB basic elements like multipliers, adders and delays. Multipliers usually have the highest implementation or computational cost and thus it is desired to reduce the number of multipliers in different systems. For each unit delay, delays can be realized by providing a storage register. After synthesis of the above digital filters in FPGA Spartan-6, the dedicated hardware resources utilization is summarized in Table 12.2 to Table 12.5, which show reduced multipliers. Since multiplication is an operation which requires large chip area and more power consumption due to repeated addition, reduction in its number helps. But the inherent property of the Spartan-6 FPGA architecture having 132 slices of DSP48A1, which supports many functions of that of an 18×18 bit multiplier, MACs, Pre-adders/subtractors (User Guide UG-389 (v1.2) 2014), wide bus multiplexers, and magnitude comparator/wide counter reduces the number of multipliers and adders utilized, suggesting small chip area and low power consumptions. The inbuilt basic structure of MAC unit using pipeline registers between multipliers and accumulator also increases the throughput as reported by a study [4]. Therefore, the FPGA implementations of digital filters design require less number of operators as compared to implementation in MATLAB environment. The number of multipliers has been reduced by traditional FPGA implementation approach [24].

To make the FPGA an ideal fit and viable alternative in various market applications, the final product can further be made attractive by optimizing the hardware components to be accommodated in a small chip area for low power consumptions. It should be remembered that the study can be further extended to synthesize the optimized digital filter design for denoising ECG signal on FPGA. This can further be applied on other available techniques and methods like ant colony optimization (ACO), genetic algorithm (GA), particle swarm optimization (PSO), etc. It is expected to get far better results in respect to either speed or area.

References

1. Farooq, U., Marrakchi, Z., & Mehrez, H. (2012). Tree-based heterogeneous FPGA architectures: application specific exploration and optimization. *Springer Science & Business Media.*
2. Kolawole, E. S., Ali, W. H., Cofie, P., Fuller, J., Tolliver, C., & Obiomon, P. (2015). Design and Implementation of low-pass, high-pass and band-pass finite impulse response (FIR) filters using FPGA. *Circuits and Systems*, 6(02), 30-48.
3. Bhogeshwar, S. S., Soni, M. K., & Bansal, D. (2019). Study of structural complexity of optimal order digital filters for de-noising ECG signal. *International Journal of Biomedical Engineering and Technology*, 29(2), 101-133.
4. Chou, C. J., Mohanakrishnan, S., & Evans, J. B. (1993). FPGA implementation of digital filters. In *Proceeding of ICSPAT* (Vol. 93, p. 1).
5. Kuon, I., Tessier, R., & Rose, J. (2008). FPGA architecture: Survey and challenges. *Foundations and Trends® in Electronic Design Automation*, 2(2), 135-253.
6. Ravikumar, M. (2012). Electrocardiogram signal processing on FPGA for emerging healthcare applications. *International Journal of Electronics Signals and Systems*, 1(3), 91-96.
7. Islam, S. M. R., Sarker, R., Saha, S., & Uddin, A. N. (2012, May). Design of a programmable digital IIR filter based on FPGA. In *2012 International Conference on Informatics, Electronics & Vision (ICIEV)* (pp. 716-721). IEEE.
8. Dixit, H. V., & Gupta, D. V. (2012). IIR filters using Xilinx System Generator for FPGA implementation. *International Journal of Engineering Research and Applications*, 2(5), 303-307.
9. Kasetwar, A. R., & Gulhane, S. M. (2013). Adaptive Power Line Interference Canceller: A Survey. *International Journal of Advances in Engineering & Technology*, 5(2), 319.
10. Pawar, D. J., & Bhaskar, P. C. (2013). FPGA based FIR filter design for enhancement of ECG signal by minimizing base-line drift interference. *International Journal of Current Engineering and Technology*, 3(5), 1775-1778.
11. Yadav, S. K., & Mehra, R. (2014). Analysis of FPGA based recursive filter using optimization techniques for high throughput. *International Journal of Engineering and Advanced Technology*, 3, 341-343.
12. Narsale, R. M., Gawali, D., & Kulkarni, A. (2014). FPGA Based Design & Implementation of Low Power FIR Filter for ECG Signal Processing. *International Journal of Science, Engineering and Technology Research (IJSETR)*, 3(6), 1673-1678.
13. Jairath, A., Shah, S.K., & Jain, A. (2012). Design & Implementation of FPGA Based Digital Filters. *International Journal of Advanced Research in Computer Engineering & Technology*, 1(7), 199-202.
14. Vijaya, V., Baradwaj, V., & Guggilla, J. (2012). Low power FPGA implementation of real-time QRS detection algorithm. *International Journal of Science, Engineering and Technology research*, 1(5), 140-144.
15. Singh, H. P., Sarin, R., & Singh, S. (2011). Implementation of high speed FIR filter using serial and parallel distributed arithmetic algorithm. *International Journal of Computer Applications*, 25(7), 26-32.
16. Gaikwad, P. K. (2013). FPGA Based Hardware Level Analysis of Inverse Sinc Filters. *International Journal of Computer Science and Mobile Applications*, 1(3), 35-39.

17. Kumar, P. R., & Meduri, M. (2013). The Implementation Of High Order Matched Fir Filter With Distributed Arithemetic. *International Journal of VLSI and Embedded Systems*, 4, 673-676.

18. Moschetta, J. M. (2014). The aerodynamics of micro air vehicles: technical challenges and scientific issues. *Internation Journal of Engineering Systems Modelling and Simulation*, 6(3/4), 134-148.

19. Meftah, S., Barbe, F., Taleb, L., & Sidoroff, F. (2007). Parametric numerical simulations of TRIP and its interaction with classical plasticity in martensitic transformation. *European Journal of Mechanics-A/Solids*, 26(4), 688-700.

20. Singh, S., & Bansal, D. (2016). A real-time acoustic signature-based fluid identification methodology for applications in the field of security and defence. *International Journal of Engineering Systems Modelling and Simulation*, 8(4), 273-283.

21. Zhang, X., & Yao, S. (2015). Fuzzy stochastic Petri nets and analysis of the reliability of multi-state systems. *IET Software*, 9(3), 83-93.

22. Wang, F., Ding, C., Zhang, L., Liu, J., & Li, R. (2012). Fetal electrocardiogram extraction algorithm in noise: using BSE. *International Journal of Biomedical Engineering and Technology*, 10(2), 199-209.

23. Bhogeshwar, S. S., Soni, M. K. and Bansal, D. (2014) "To verify and compare de- noising of ECG signal using various de-noising algorithms of IIR & FIR filters", International Journal of Biomedical and Engineering Technology, INDERSCIENCE, Scopus (Elsevier), Volume.16, Issue. 3, 2014, pp:244-267.

24. Bhogeshwar, S. S., Soni, M. K., & Bansal, D. (2015). Circuit system analysis for real-time acquisition of bio-signals. *International Journal of Biomedical Engineering and Technology*, 18(3), 272-289.

25. Nayak, S., & Rai, A. (2019, December). Synthesis and Analysis of Optimal Order Butterworth Filter for Denoising ECG Signal on FPGA. In *International Conference on Information Management & Machine Intelligence* (pp. 359-369). Springer, Singapore.

26. Bokde, P. R., & Choudhari, N. K. (2015). Implementation of digital filter on FPGA For ECG signal processing. *International Journal of Emerging Technology and Innovative Engineering*, 1(2), 175-181.

27. Hrairi, M., & Baharom, B. H. B. (2013). Design and modelling of silicon MEMS accelerometer. *International Journal of Engineering Systems Modelling and Simulation*, 5(4), 181-187.

28. Martini, N., Milanesi, M., Vanello, N., Positano, V., Santarelli, M., & Landini, L. (2010). A real-time adaptive filtering approach to motion artefacts removal from ECG signals. *International Journal of Biomedical Engineering and Technology*, 3(3-4), 233-245.

29. Singh, S., & Bansal, D. (2016). A real-time acoustic signature-based fluid identification methodology for applications in the field of security and defence. *International Journal of Engineering Systems Modelling and Simulation*, 8(4), 273-283.

30. Thakur, R., & Khare, K. (2013). High speed FPGA implementation of FIR filter for DSP applications. *International Journal of Modeling and Optimization*, 3(1), 92-94.

31. Wang, F., Ding, C., Zhang, L., Liu, J., & Li, R. (2012). Fetal electrocardiogram extraction algorithm in noise: using BSE. *International Journal of Biomedical Engineering and Technology*, 10(2), 199-209.

32. Xilinx Inc., *Synthesis and simulation Design Guide*, June 2008.

33. Xilinx Spartan-6 Family Overview, DS160 (v2.0) October 25, 2011. [25Xilinx XPower Estimator User GuideUG440 (v13.4) January 18,2012.

34. Nayyar, A., Le, D. N., & Nguyen, N. G. (Eds.). (2018). *Advances in swarm intelligence for optimizing problems in computer science*. CRC press.

35. Zhang, X., & Yao, S. (2016). Multi-state systems reliability with composite importance measures of fuzzy petri nets. *International Journal of Engineering Systems Modelling and Simulation*, 8(4), 255-263.

13

Neural Networks and Their Applications

Shivani Joshi[1], Anju Gera[1], Sweta Bhadra[2]

[1] GL Bajaj Institute of Technology and Management, India
[2] Assam Down Town University, India
Email: shivani1275@gmail.com, anju.gera@gmail.com, sweta.bhadra4@gmail.com

Abstract
Neural networks are associated with human brains as the human brain consists of 10,000 billion nerves. These nerves consist of neurons, which have some weight and receive signals, and these signals are processed and converted to the desired output. Similarly, neural networks are produced as a parallel gadget, which can perform computational assignments quicker than the regular system. The fundamental errand of neuron networks is to perceive examples and grouped capacities dependent on the estimation, improvement, and information bunching. So, neural networks are known as artificial neural network (ANN). This network fills in as a human cerebrum and age endeavours to take care of the overwhelming issues. ANN is parallel distributed processing systems or connective systems. In this chapter, we discuss the basic working of neurons, compare and contrast biological neural networks (BNN) with artificial neural networks (ANN), and discuss different types of learning systems used in ANN. We also discuss various applications of neural networks in real life.

Keywords: Artificial neural network (ANN), biological neural network (BNN), network topology, feedforward network (FFN), computer-aided tracking and characterization of homicides (CATCH)

13.1 Introduction

In our brain, there are ten billion cells, which are correspondingly called neurons, and they process the information as electric signals. As shown in Figure 13.1, each neuron is synoptically connected to another neuron, and the dendrites of each neuron carry external information that enters the system. Each association is connected to a load that holds signals containing information. Depending on the quality of the signal, the neuron can accept or reject the data. Since the load consists of a signal sent across distinct neurons, the primary task of neurons is to handle a specific problem. An activation signal is a type of internal expression found in every neuron. Output or desired motions are produced when information signals and the actuation rule for multiple components are coupled [1,2].

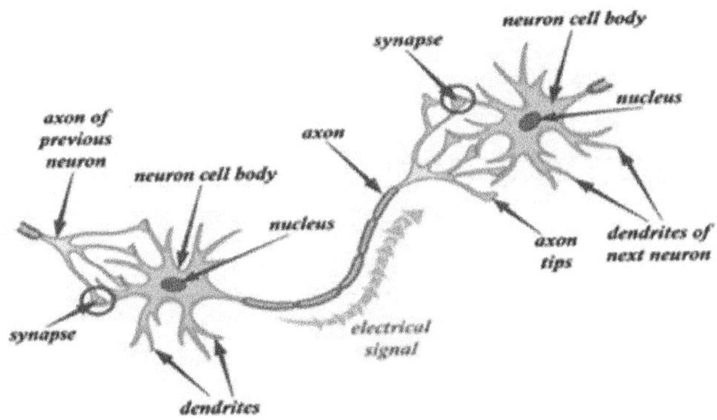

Figure 13.1: Structure of the neuron network.

13.2 Main Work of Neuron

The neuron consists of four parts, and each part performs its work, which is described below [3]:

a) Dendrites: Their structure is tree-like with branches. Their work is to receive the information from neurons to which they are connected.

b) Soma: This is the cell form of a neuron which processes the information acquired from dendrites.

c) Axon: It resembles a link, which fills in as an intermediate over which neurons refer the data.

d) Synapse: The junction between the axon of one neuron and the dendrite of another.

13.3 Comparison Between Artificial Neural Network (ANN) and Biological Neural Network (BNN)

Let us discuss the similarities and differences between ANN and BNN. Table 13.1 shows the similarities between ANN and BNN [4]:

Table 13.2 shows the differences between ANN and BNN [4]:

Table 13.1: The similarities between ANN and BNN.

Biological Neural Network (BNN)	Artificial Neural Network (ANN)
Soma	Node
Dendrites	Input
Synapse	Weights or Interconnections
Axon	Output

Table 13.2: The differences between ANN and BNN.

Criteria	BNN	ANN
Processing	It is parallel and has superior processing as compared to ANN.	It is parallel and has fast processing but less than BNN.
Size	The size is 10^{11} neurons and 10^{15} interconnections.	The size is 10^2 to 10^4 nodes (mainly be determined by on the type of application and network designer).
Learning	It is able to tolerate uncertainty.	They are specific and organized. The arranged information is required to stand uncertainty.
Fault tolerance	The enactmentdestroysthrough small damage.	Thisis a stout performer and has the perspective to be liability accepting.
Storage capacity	It stocks every info in the synapse	Info is stored in contiguous memory sites

13.4 How Artificial Neural Network Works

Figure 13.2 shows the working model of artificial neural network.

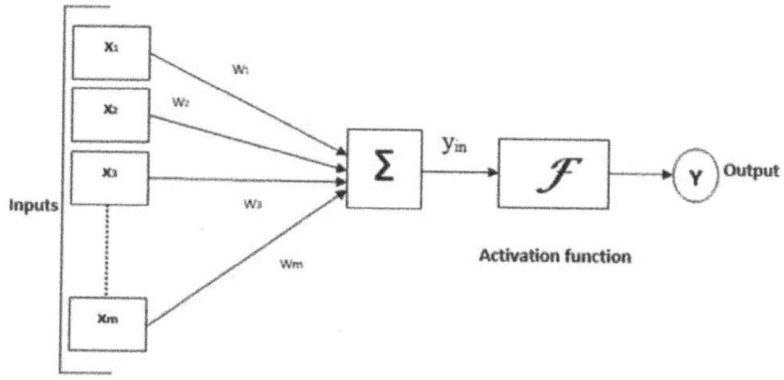

Figure 13.2: Working model of artificial neural network.

Here, $W_1, W_2, W_3, ..., W_m$ are the strength of the input signals.
As per Figure 13.2, the net input can be determined as:

$$Y_{in} = X_1W_1 + X_2W_2 + X_3W_3 + ... + X_mW_m \quad (13.1)$$

i.e., Net input $Y_{in} = \sum_i^m X_i W_i$.
The output can be determined by using the activation function of the net input.

$$Y_{out} = F(Y_{in}) \tag{13.2}$$

$$Output = Function(NetInputCalculated) \tag{13.3}$$

13.4.1 Processing of ANN Building Blocks

There are three main building blocks of ANN used for processing:

- Network Topology
- Adjustments of Weights or Learning
- Activation Function

13.4.1.1 Network Topology

A network topology is the physical and logical arrangement of nodes and connections in a network. It is classified into three networks according to network topology of ANN:

- Feedforward Network (FFN): Feedforward networks are non-intermittent systems with layers of handling hubs. These hubs within a layer are linked to the hubs inside earlier levels. Each association has a different weight. Usually, the signal flows in a single direction in input-to-output form. It is also isolated into two sorts [5].
  - Single layer feedforward network: It is a single weighted layer where the information layer is completely associated with the output-giving layer, as shown in Figure 13.3.

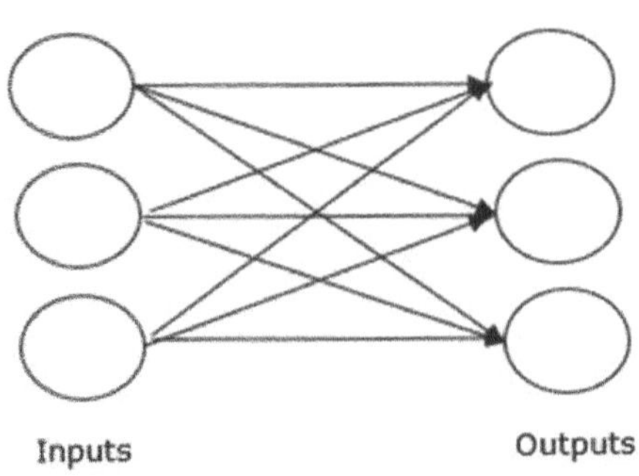

Figure 13.3: Single layer feedforward network.

  - Multilayer feedforward network: There is at least one layer between the information and the generous layer. These are called hidden or concealed layers, as shown in Figure 13.4.

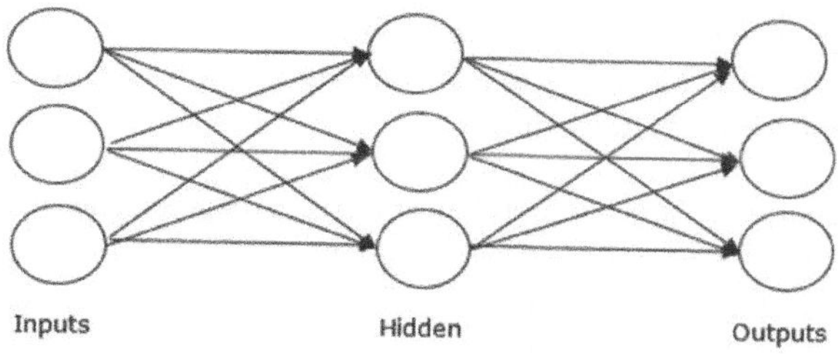

Figure 13.4: Multilayer feedforward network.

This system has feedback response routes, which means that signals flow in both directions in loops. The feedback network has nonlinear properties with the dynamic design and continually evolves until it finds equilibrium. It is also classified into three types [6].

- Recurrent networks: These types of feedback systems have closed loops. There are two kinds of recurrent networks.
- Fully recurrent network: This type of network has a simple neural system structural design since all nodes are related nodes, and each node works as both input and output as shown in Figure 13.5.

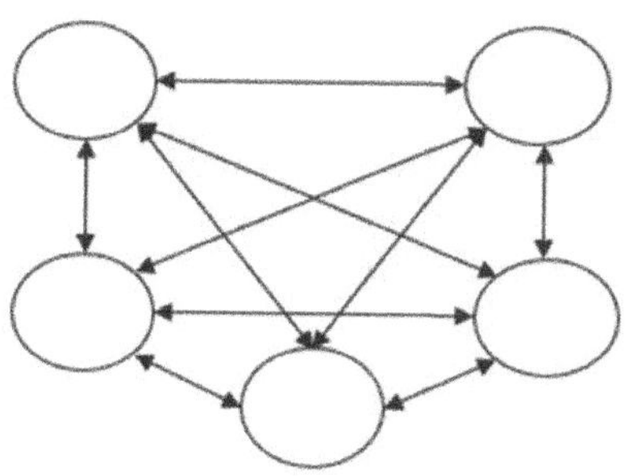

Figure 13.5: Fully recurrent network.

- Jordan network: This is a closed loop network where input and output serve in such a way that one output will work as the input for another node (see Figure 13.6).

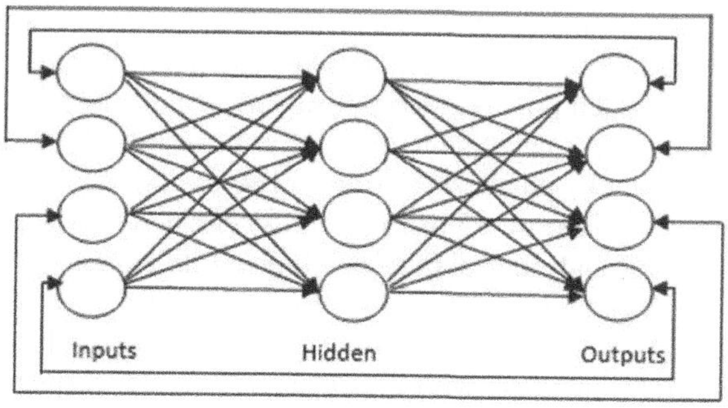

Figure 13.6: Jordan network.

13.4.1.2 *Adjustments of Weights or Learning*

This is a technique for changing many connections between the neurons in a certain system. There are three ways to learn:

- Supervised Learning: It is a dependent process, hence executed under supervision. In this learning, the info vector behaves as input to the system and is trained to give an output of the desired vector. The vector selected will be contrasted with the normally given vector. Based on this, error signals are produced. The weights are adjusted until and unless the actual output is not verified with the expected one [7] (see Figure 13.7).

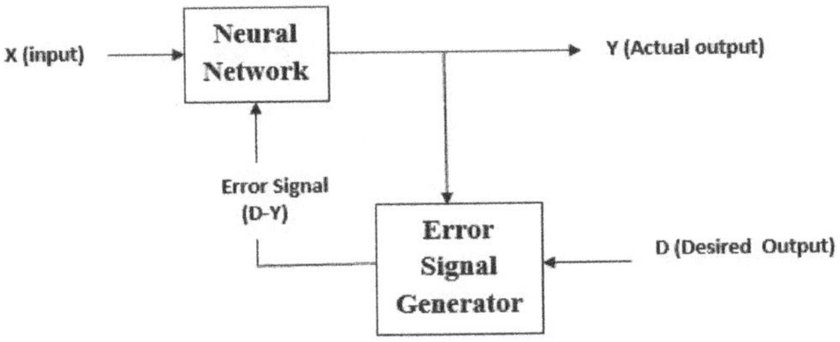

Figure 13.7: Supervised learning.

- Unsupervised Learning: This is managed without supervision, and the learning process is autonomous. In this learning, the cluster is being produced by combining similar types of input vectors. The creation of the desired output vector is based on the learning of the input vector. The input data in this network determines the new learning patterns and features, which depict the link between input and output [8,9] (see Figure 13.8).

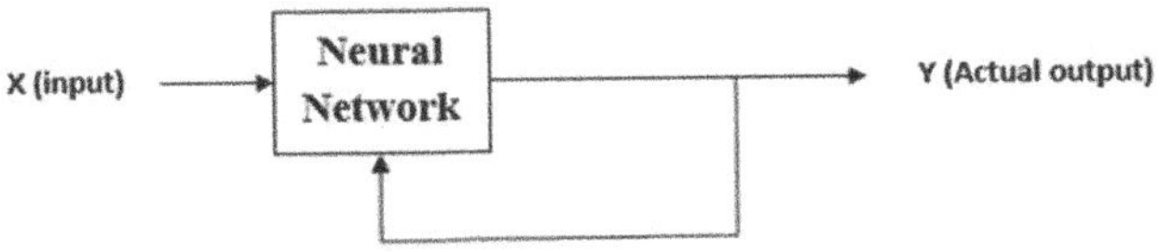

Figure 13.8: Unsupervised learning.

- Reinforcement Learning: In this type of learning, an agent may gain knowledge by trial and error while receiving feedback from its actions and experiences in an interactive environment. Each positive step praises the agent, whereas each negative action results in punishment or negative feedback for the agent. The agent interacts with the surroundings and looks into it on its own. The basic goal of an agent in reinforcement learning is to maximize positive reinforcement while performing better. The agent learns via hit-and-miss and, depending on its experience, develops the skills necessary to carry out the mission more effectively. Thus, it can be said that "Reinforcement learning is a form of machine learning approach where an intelligent agent (computer program) interacts with the environment and learns to function within it." A robot dog's ability to learn how to move its limbs is an illustration of reinforcement learning [10,11] (see Figure 13.9).

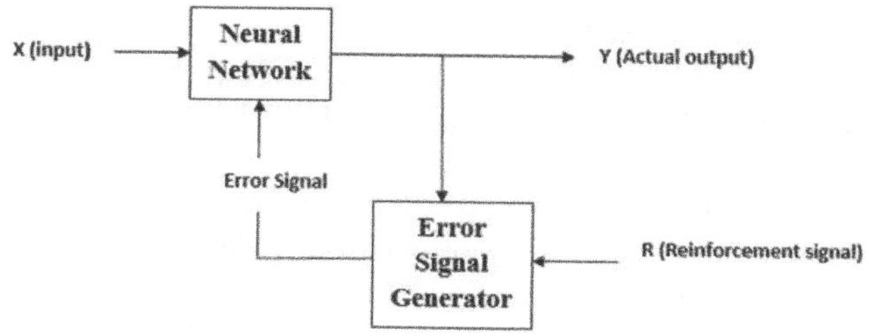

Figure 13.9: Reinforcement learning.

13.4.1.3 Activation Function

The enacting capacity refers to the additional effort or energy required to get an accurate result [12,13]. In ANN, enactment capacities are related to the contribution to produce the desired output. There are two capabilities for enactment, which are as follows:

1. Linear Activation Function: This function is also known as an identity function because it performs having no input editing [14]. It is defined by:

$$F(x) = x \tag{13.4}$$

2. Sigmoid Activation Function [15]:

   a) Binary sigmoidal function: A logistic function with binary output values ranging from 0 to 1 is known as a "binary sigmoid function." It generates non-binary activations, is nonlinear, and is differentiable. However, Sigmund's vanishing gradients are a concern. Additionally, the function of sigmoid activation is not zero-centric

$$F(x) = sigm(x) = \frac{1}{1 + exp(-x))} \tag{13.5}$$

   b) Bipolar sigmoidal function: A popular activation function or "squashing function" in Hopfield neural networks is the bipolar sigmoid. The sigmoid function and this function are linked. Since it is a bounded function, this activation function works well for programs that generate output values between $[-1, 1]$. Additionally, the activity of each neuron ranges from -1 to 1. Compared to the McCulloch-Pitts function, this function may be a better option for training the data between 0 and 1. Consequently, it is simpler to distinguish between bipolar and monotonic sigmoid activation functions.

$$F(x) = sigma(x) = \frac{2}{1 + exp(-x)} - 1 = \frac{1 - exp(x)}{1 + exp(x)} \tag{13.6}$$

13.4.2 Neural Network Learning Rules

A learning rule or learning process is a method or a mathematical logic. It improves the artificial neural network and uses this rule on the whole network. Learning rules change a network's weights and bias levels when it simulates in a certain data environment. It is a way to use a learning rule that goes back and forth. It lets a neural network learn from its environment and get better at what it does. Here are the different types of rules for how neural networks learn:

13.4.2.1 Hebbian Learning Rule

This is one of the oldest and simplest rules. Donald Hebb presented this rule in his book *The Organization of Behavior* published in 1949. Hebb gives the concept of the rule, stating that: "When an axon of cell A is near enough to excite a cell B and repeatedly or persistently takes part in firing it, some growth process or metabolic change takes place in one or both cells such that A's efficiency, as one of the cells firing B, is increased" [16].

Starting with the statement, we can assume and determine that the links between two neurons can become strong if the neurons fire at the same time and can become weak if neurons fire at different times [16,17].

Mathematical formulation: According to the Hebbian learning rule, the weight adjustment function [18] at each time is shown in Equation (13.7).

$$\Delta W_{ji}(t) = \alpha x_i(t) y_j(t) \tag{13.7}$$

where,

- $\Delta W_{ji}(t)$ = increment that weight of linking increases at time step *t*;
- $\alpha =$ the positive as well as constant learning rate;
- $\alpha x_i(t) =$ input value of pre-synaptic neuron at time step *t*; and
- $y_j(t) =$ output of pre-synaptic neuron at same time step *t*.

13.4.2.2 Perceptron Learning Rule

In neural network connections, each has a weight that varies as the network learns. It is an illustration of supervised learning, in which the network begins its learning process by giving each weight a random value. Determining the output value based on a collection of records for which we know the predicted output value for a sample is called the learning rule. The network then compares the estimated output value to the anticipated value. The sum of the squares of the errors that occurred for each person in the learning sample can be used to build an error function, which is the next step [19].

Mathematical formulation: Let '*n*' be the number of finite input vectors, $x(n)$ expected through the desired output vector, which is desired $t(n)$, where $n = 1$ to *N* [19,20].

We can calculate the output '*Y*', by the net input, as well as the activation function as follows :

$$y = f(y_{in}) = \begin{cases} 1, & y_{in} > \theta \\ 1, & y_{in} \leq \theta \end{cases} \tag{13.8}$$

where θ is a threshold.

The entire process's weight adjustment can be obtained using these two cases:

A. Case I - when $t \neq y$, then
   - $W(new) = W(old) + tx$
   - $W(new) = W(old) + tx$

B. Case II - when $t = y$, then
   - $W(new) = W(old)$

13.4.2.3 Delta Learning Rule (Widrow-Hoff Rule)

This rule is called the least mean square strategy (LMS) invented by Bernard Widrow and Marcian Hoff. This standard reduces the number of preparation designs that go wrong. It is a learning calculation that requires constant application work. Its main standard concept is an angle drop technique. The changes in the synaptic loads limit the given unit to the objective esteem's net contribution [21].

Mathematical Formulation: The mathematical representation of this rule is to update the synaptic weights that are given by:

$$\Delta W_i = \alpha x_i e_j \tag{13.9}$$

where

- ΔW_i = change in weight for the ith outline;
- α = positive as well as constant learning rate;
- x_i = input significance from pre-synaptic neuron;
- $e_j = (t - Y_{in})$, difference among target output as well as actual output Y_{in}.

In every case, we get the single output by adjusting weights differently.

1. Case I - when $t \neq y$, then $W(new) = W(old) + \Delta w$
2. Case II - when $t = y$, then No change in weight

13.4.2.4 Competitive Learning Rule (Winner-Take-All)

This rule works with unsupervised learning, in which similar input vectors are grouped together. When a new pattern is given to the neural network as input, it gives back an output that tells the pattern's class. It does not care if the environment tells whether the desired output is right or wrong. In this rule, the winner is the neuron with the most inputs. The connections between the output neurons are the competitors. The "ON" connection is the winner, while the "OFF" connections are the loser [22] (see Figure 13.10).

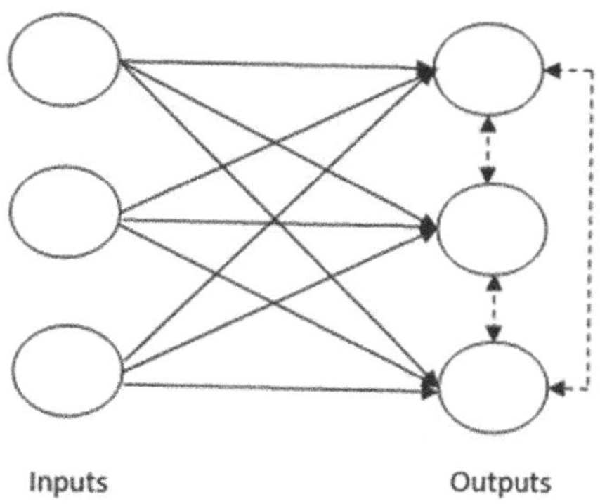

Figure 13.10: Concept of competitive network.

We have already said that the output nodes will compete with each other. The main idea is that the winner will be the output unit that reacts most strongly to a certain input pattern during training. This rule is also called "winner-takes-all" because only the neuron that wins is changed. The rest of the nerve cells do not change [23].

Mathematical formation: There are three most important things that mathematicians need to know about this learning rule.

1. Condition of a winner: If a neuron y_k wants to win, it must meet the following requirements:

$$y_k = \begin{cases} 1, & \textit{if } v_k > v_j \textit{ for all } i, j \neq k \\ 0, & , \textit{ otherwise} \end{cases} \tag{13.10}$$

   It means that if a neuron, say y_k, wants to win, its local induced field, which is the output of the summation unit v_k, must be the biggest among all the other neurons in the network.

2. Condition of the total of weight: The competitive learning rule is also limited by the fact that the sum of the weights for a given output neuron will always be 1. For example, if we think about neuron *k*, we can say:

$$\sum_j w_{kj} = 1 \ \forall k \tag{13.11}$$

3. Change of weight for winner: If a neuron does not respond to the input pattern, that neuron does not learn anything. However, if a certain neuron wins, the weights of the other neurons are changed as in:

$$\Delta w_{kj} = \begin{cases} -\alpha(x_j - w_{kj}), & \textit{if neuron k wins} \\ 0, & , \textit{ if neuron k losses} \end{cases} \tag{13.12}$$

   where α is the learning rate.

Here, it is clearly mentioned that the winning neuron is favored by regulating its weight. Otherwise,there is a neuron loss; it need not be a problem to re-adjust its weight.

13.4.2.5 Outstar Learning Rule

Grossberg came up with this rule, which is an example of supervised learning, as the goals are clear. The rule is put into place when nodes in a network are set up in layers. Here, the weights that are connected to a certain node should be the same as the outputs. It was made to give the layer of *p* neurons the output *d* that was wanted [22].

Mathematical Construction: The weight adjustments by:

$$\Delta w_j = \alpha(d - w_j) \tag{13.13}$$

Where *d* is the desired neuron output, and α is the learning rate.

13.5 Neural Networks and Their Applications

The ANN model is now widely used in many applications to identify and solve problems. It is a great model that is used in a lot of different areas, such as for early detection of diseases so that proper treatment can be done right away, telemedicine in rural areas, security framework, account management part, securities exchange, agribusiness and safety framework [24]. Some of the applications have been made clear.

13.5.1 Image Processing and Character Recognition

Neural networks help to recognize characters and images. It helps to solve the problem of bank fraud by recognizing handwriting and even national security assessments. ANN does this by taking in much information and processing it to find hidden, complex, and nonlinear relationships. Image recognition or the processing is done from facial recognition in social media. Face recognition divides the image into two parts, containing targets (faces) and providing the background. Then, the image is analyzed, and the faces are identified. Image processing is also used in the agriculture and defense sectors. Image processing is also used in the medical area to detect cancer. In this context, ANN refers to deep neural networks, which are the foundation of deep learning. Deep learning has facilitated revolutionary advances in computer vision, speech recognition, and natural language processing [25].

13.5.2 Business Forecasting

In the modern world, business influences the economy, and the neural network is utilized for business forecasting to get the right flow of information and make wise business decisions. The distribution of funds among products is a common sales practice. Utilizing capacity requires consideration of financial and stock market economic and monetary policies. The stock market is complicated, making price predictions quite difficult. To overcome the constraints of standard forecasting models, which are complicated and rely on nonlinear correlations, all hidden and underlying aspects must be considered. The neural network is a better alternative as it contains relationships and model-abstract qualities. The classic models have no restrictions whatsoever on the input and residual distributions. Recent developments in using LSTM and recurrent neural networks for forecasting are the subject of more investigations [26,27].

13.5.3 Financial Prediction

Financial prediction is an effective instrument for stock market forecasting. It is astonishing how well firms perform when employing the neural network prediction method, as evidenced by MJ Futures' two-year return of 199.2%. You can skip developing complex rules (and redeveloping them as their effectiveness fades), says technical editor John Sweeney in the 1995 issue of "Technical Analysis of Stocks and Commodities.". Just define the price series and indicators you want to use, and the neural network takes care of the rest. These examples show that neural networks may be simple to use once they are set up, but setting them up and getting them ready takes skill, engagement, and effort. Neural networks can identify patterns in data that humans are unlikely to notice and use those patterns to make accurate projections.

A perfect example of this is given by Dean Barr and Walter Loick of LBS Capital Management, using only six financial points as data sources in a neural network that is generally simple. These data sources are the ADX, which shows the typical directional development over the last eighteen days, and the current estimate of the S&P 500, which shows the significant change in the S&P 500 incentive five days earlier [26,28].

13.5.4 Additional Neural Network Uses in the Economic World

Additional neural network uses in the economic world include:

- Currency prediction;

- Futures prediction;
- Bond ratings;
- Business failure prediction;
- Debt risk assessment;
- Credit approval;
- Bank theft.

13.5.5 The Traveling Salesman Problem

The traveling salesman problem (TSP) is the challenge of finding the shortest, most efficient route for a person to take, given a list of specific destinations. It is a well-known algorithmic problem in the fields of computer science and operations research, with important real-world applications for logistics and delivery businesses [29]. There are obviously a lot of different routes to choose from, but finding the best one — the one that will require the least distance or cost — is what mathematicians and computer scientists have spent decades trying to solve. It's much more than just an academic problem. Finding more efficient routes, or route optimization, increases profitability for delivery businesses, and reduces greenhouse gas emissions because it means less distance traveled.

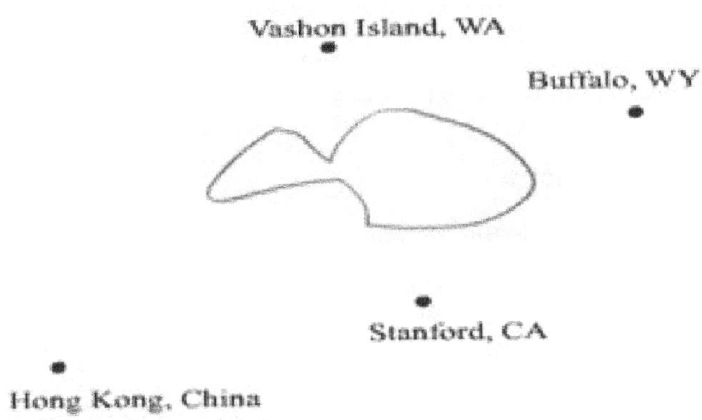

Figure 13.11: Elastic band for the shortest path.

Example: A list of all the cities is provided to help with a plan to determine the shortest path between two points that allows one to visit every town. It employs the elastic net with random orientation and the recursion approach for each opportunity. To organize a Kohonen SOM (self-organizing map) like an elastic rubber band, we can use it to obtain an approximation of the answer.

The algorithm is used to display a different random city every time and the next nearest point on the internet to show the closest city [30]:

1. Because the net is elastic, as shown in Figure 13.11, it changes shape as people pull on it. Due to the elasticity of the ring, the length of the ring tends to shorten when various people pull the net coverings into a ring for the various towns around it. Thus, the TSP can be roughly estimated using the approach [47].

2. The weight vectors of the networks are initially randomly assigned in SOMs by employing simple competitive networks. When an input vector is provided, SOM's other relationship is to determine the winner (Perceptron, whose weight vector is closest to the input vector) [31].

3. The Kohonen SOM is used to solve the TSP [32]:

4. It replicates an elastic band and treats the load vectors as focused on a plane because the flexible elastic band has a ring-shaped structure.

5. The probability that a town's directions are x and y are introduced as the system's information vector. It mixes these attentions by the position of their perceptrons in the ring with the best layer. The system detects the load vector closest to the town and adjusts toward it and its neighbors. Therefore, load vectors behave like the focus of an elastic band. In this way, each lesson is comparable to pulling the band's closest goal in the direction of a town. According to the standard, the elastic band's elastic property causes the measure of learning to alter in opposition to the physical separation between the hub and the vector.

13.5.6 Medicine

Neural networks diagnose breathing and heart problems in different ways, such as with auscultation coats, test accounts, and precordial phonocardiograms (PCG). A patient may have regular checkups in an area with a higher chance of finding a disease or injury [33].

For different models, the records or information may include blood pressure, heart rate, breathing rate, ECG, ICG, etc. The models must show how people of different ages, genders, and sizes move. Physiological information from the present and the past, as well as information from the other conventional models, are compared. Then, deviations from this standard are compared to what is known to be the real cause of each medical illness. Also, neural networks can learn by focusing on different conditions and models and combining them to make a total calculated representation that can be used to figure out a patient's disease based on the models [34].

13.5.7 Electronic Nose

An electronic nose is an electronic sensing device intended to help computers find and describe smells, gases, and vapors. It consists of a system for detecting substances, like a spectrometer, that can pick up on the specific example of synthetic mixtures. When the concoction sensor cluster detects a smell, these synthetics are turned into a format that the computer can understand. The fake neural system recognizes the ingredient list used in different fields, such as the health, medical, and food industries [35,36]. Some of the applicative areas for an electronic nose are:

a) Environment: It can find toxic waste, analyze fuel blends and gases, find oil holes and tell the difference between proof of family smells. It can also check air quality, prevent industrial facility emissions, and test groundwater for smells.

b) Medical: It is used in the medical field to look at smells coming from the body to find and diagnose problems. Problems can be shown by scents in the air, contaminated wounds, and body fluids. It also discovers tuberculosis with the help of neural networks.

c) Food: Food is the most important practical market in the food industry. An electronic nose helps to check the food quality, checks for rancidity in mayonnaise, programs flavor control, and monitors the aging process of cheddar cheese. It also checks if the juices used are common, surveys the refreshment compartment, and reviews bourbon.

In all the above cases, the focused neural system is quantified by the learning vector to find the metal oxide. Gas sensors are used to figure out different smells.

13.5.8 Security

The CATCH (Computer-aided tracking and characterization of homicides) program is utilized to distinguish current wrongdoing, the area of the wrongdoing, and the specific attributes of the offense. The program is additionally subdivided into various devices, and each device has a certain trademark or group of qualities. This enables the client to remove explicit attributes that people are not familiar with or decided upon [37,38].

13.5.9 Loans and Credit Cards

Banks offer services like loans and credit cards. However, from the records, we can see that many people cannot repay loans, and banks have difficulty getting the money back. Neural networks help banks decide who to lend money to and who to reject, depending on the number of times they failed to repay a loan. The neural network systems help banks [39-49] in the following ways:

a) A neural network uses past information to decide the terms and interest rate of a loan for a specific person. Banks use these rules so that they fail less often. Even some credit card companies use neural networks to decide whether or not to give someone a credit card and approve their application based on their credit score [41].

b) A bank's decision about whether or not to give a candidate credit depends on the candidate's past repayment record in order to avoid creating problems. For example, the bank or credit organization must prove to the candidate that their choice is valid. It is hard to tell a person whose credit application was turned down what the computer or neural network learned and how it made its decision.

13.5.10 Other Applications of Neural Networks

- Composing Music: Computer and other electronic devices learn the patterns in the composition used in music by deep neural nets [43].

- Robot Navigation: Robots are trained to perform human tasks by using artificial neural nets [9].

- Autonomous Driving Cars: Neural networks tell the difference between objects, people, and road signs. They can also train a vehicle to drive itself without a person. The concept behind this is a single hidden-level backpropagation network based on pictures of the road in different situations and the right steering adjustment for each situation [44-50].

13.6 Conclusion and Future Scope

In order to properly learn and understand anything, we must do so in order to put the knowledge we gain from it to use in the future. We must research the "fundamentals" of things. Artificial neural network principles provide a quick explanation of ANN. The practice of neural networking calls for much effort as data is input into the system and performances are observed, processes adjusted, connections made, and rules modified until the network reaches the desired state, which is then updated and the results repeated. These intended outcomes have a statistical basis. The network isn't always accurate. It is because of this that neural networks are appearing in applications where people are likewise incapable of always being correct. Currently, neural networks are recruiting new customers, approving loans, denying credit cards, choosing stocks, prospecting, approving and adjusting control mechanisms and reviewing output. But the future promises are even greater. Neural networks demand more rapid hardware. They must integrate into hybrid systems, which also make use of expert systems and fuzzy logic. Consequently, these systems will be able to read handwriting, hear conversation, and make action plans. They'll be able to develop into intelligent entities. Besides, machines never get tired or distracted. Thus, due to innovations, in time "intelligent" machines will be at the forefront.

References

1. Lisboa, P. J. (2002). A review of evidence of health benefit from artificial neural networks in medical intervention. *Neural Networks*, 15(1), 11-39.
2. North, B. V., Curtis, D., Cassell, P. G., Hitman, G. A., & Sham, P. C. (2003). Assessing optimal neural network architecture for identifying disease-associated multi-marker genotypes using a permutation test, and application to Calpain 10 polymorphisms associated with diabetes. *Annals of Human Genetics*, 67(4), 348-356.
3. Camps-Valls, G., & Guerrero-Martínez, J. F. (2006). Neural Networks in ECG Classification: What is Next for Adaptive Systems?. In *Neural Networks in Healthcare: Potential and Challenges* (pp. 81-104). IGI Global.
4. McClelland, J. L., Rumelhart, D. E., & PDP Research Group. (1987). Parallel Distributed Processing, Volume 2: *Explorations in the Microstructure of Cognition: Psychological and Biological Models* (Vol. 2). MIT press.
5. Svozil, D., Kvasnicka, V., & Pospichal, J. (1997). Introduction to multi-layer feed-forward neural networks. *Chemometrics and Intelligent Laboratory Systems*, 39(1), 43-62.
6. Ge, S. S., Hang, C. C., & Zhang, T. (1999). Adaptive neural network control of nonlinear systems by state and output feedback. *IEEE Transactions on Systems, Man, and Cybernetics*, Part B (Cybernetics), 29(6), 818-828.
7. Dietterich, T. G. (1998). Approximate statistical tests for comparing supervised classification learning algorithms. *Neural Computation*, 10(7), 1895-1923.
8. Kosko, B. A. (1991). Structural stability of unsupervised learning in feedback neural networks. *IEEE Transactions on Automatic Control*, 36(7), 785-792.
9. Lagerholm, M., Peterson, C., Braccini, G., Edenbrandt, L., & Sornmo, L. (2000). Clustering ECG complexes using Hermite functions and self-organizing maps. *IEEE Transactions on Biomedical Engineering*, 47(7), 838-848.
10. Lewis, F. L., & Vrabie, D. (2009). Reinforcement learning and adaptive dynamic programming for feedback control. *IEEE Circuits and Systems Magazine*, 9(3), 32-50.

11. Lewis, F. L., Vrabie, D., & Vamvoudakis, K. G. (2012). Reinforcement learning and feedback control: Using natural decision methods to design optimal adaptive controllers. *IEEE Control Systems Magazine*, 32(6), 76-105.
12. Leshno, M., Lin, V. Y., Pinkus, A., & Schocken, S. (1993). Multilayer feedforward networks with a nonpolynomial activation function can approximate any function. *Neural Networks*, 6(6), 861-867.
13. Maas, A. L., Hannun, A. Y., & Ng, A. Y. (2013, June). Rectifier nonlinearities improve neural network acoustic models. In *Proc. ICML* (Vol. 30, No. 1, p. 3).
14. Vuckovic, A., Radivojevic, V., Chen, A. C., & Popovic, D. (2002). Automatic recognition of alertness and drowsiness from EEG by an artificial neural network. *Medical Engineering & Physics*, 24(5), 349-360.
15. Kalman, B. L., & Kwasny, S. C. (1992, June). Why tanh: choosing a sigmoidal function. In [Proceedings 1992] *IJCNN International Joint Conference on Neural Networks* (Vol. 4, pp. 578-581). IEEE.
16. Hagan, M. T., Demuth, H. B., & Beale, M. (1997). *Neural network design*. PWS Publishing Co., Boston.
17. Sanger, T. D. (1989). Optimal unsupervised learning in a single-layer linear feedforward neural network. *Neural Networks*, 2(6), 459-473.
18. Gerstner, W., & Kistler, W. M. (2002). Mathematical formulations of Hebbian learning. *Biological Cybernetics*, 87(5), 404-415.
19. Sussner, P. (1998, September). Morphological perceptron learning. In *Proceedings of the 1998 IEEE International Symposium on Intelligent Control (ISIC) held jointly with IEEE International Symposium on Computational Intelligence in Robotics and Automation (CIRA) Intell* (pp. 477-482). IEEE.
20. Jain, A. K., Mao, J., & Mohiuddin, K. M. (1996). Artificial neural networks: A tutorial. *Computer*, 29(3), 31-44.
21. Wilamowski, B. M. (2009). Neural network architectures and learning algorithms. *IEEE Industrial Electronics Magazine*, 3(4), 56-63.
22. DeSieno, D. (1988, July). Adding a conscience to competitive learning. In *IEEE International Conference on Neural Networks* (Vol. 1).
23. Grossberg, S. (1988). Nonlinear neural networks: Principles, mechanisms, and architectures. *Neural networks*, 1(1), 17-61.
24. Gill, S., & Pawar, H. (2013). Neural networks and its applications. *Asian Journal of Multidimensional Research (AJMR)*, 2(1), 83-89.
25. Parisi, R., Di Claudio, E. D., Lucarelli, G., & Orlandi, G. (1998, May). Car plate recognition by neural networks and image processing. In *1998 IEEE International Symposium on Circuits and Systems (ISCAS)* (Vol. 3, pp. 195-198). IEEE.
26. Karels, G. V., & Prakash, A. J. (1987). Multivariate normality and forecasting of business bankruptcy. *Journal of Business Finance & Accounting*, 14(4), 573-593.
27. Zhang, G. P. (Ed.). (2004). *Neural networks in business forecasting*. IGI Global.
28. Trippi, R. R., & Turban, E. (Eds.). (1992). *Neural networks in finance and investing: Using artificial intelligence to improve real world performance*. McGraw-Hill, Inc..
29. Angeniol, B., Vaubois, G. D. L. C., & Le Texier, J. Y. (1988). Self-organizing feature maps and the travelling salesman problem. *Neural Networks*, 1(4), 289-293.
30. Wilson, G. V., & Pawley, G. S. (1988). On the stability of the travelling salesman problem algorithm of Hopfield and Tank. *Biological Cybernetics*, 58(1), 63-70.

31. Schabauer, H., Schikuta, E., & Weishaupl, T. (2005, December). Solving very large traveling salesman problems by SOM parallelization on cluster architectures. In *Sixth International Conference on Parallel and Distributed Computing Applications and Technologies (PDCAT'05)* (pp. 954-958). IEEE.

32. Aras, N., Oommen, B. J., & Altınel, İ. K. (1999). The Kohonen network incorporating explicit statistics and its application to the travelling salesman problem. *Neural Networks*, 12(9), 1273-1284.

33. Okamura, N., Arai, H., Maruyama, M., Higuchi, M., Matsui, T., Tanji, H., ... & Sasaki, H. (2002). Combined analysis of CSF Tau levels and [123I] Iodoamphetamine SPECT in mild cognitive impairment: implications for a novel predictor of Alzheimer's disease. *American Journal of Psychiatry*, 159(3), 474-476.

34. Penny, W., & Frost, D. (1996). Neural networks in clinical medicine. *Medical Decision Making*, 16(4), 386-398.

35. Hines, E. L., Llobet, E., & Gardner, J. W. (1999). Neural network based electronic nose for apple ripeness determination. *Electronics Letters*, 35(10), 821-823.

36. Weng, H., Dong, X., Hu, X., Beetner, D. G., Hubing, T., & Wunsch, D. (2005, August). Neural network detection and identification of electronic devices based on their unintended emissions. In *2005 International Symposium on Electromagnetic Compatibility, 2005. EMC 2005.* (Vol. 1, pp. 245-249). IEEE.

37. Barr, D. S., & Mani, G. (1998). Predictive neural network means and method for selecting a portfolio of securities wherein each network has been trained using data relating to a corresponding security. U.S. Patent No. 5,761,442. Washington, *DC: U.S. Patent and Trademark Office*.

38. Niebur, D., & Germond, A. J. (1992). Power system static security assessment using the Kohonen neural network classifier. *IEEE Transactions on Power Systems*, 7(2), 865-872.

39. Akkoç, S. (2012). An empirical comparison of conventional techniques, neural networks and the three stage hybrid Adaptive Neuro Fuzzy Inference System (ANFIS) model for credit scoring analysis: The case of Turkish credit card data. *European Journal of Operational Research*, 222(1), 168-178.

40. Jensen, H. L. (1992). Using neural networks for credit scoring. *Managerial Finance*, 18 (6):15-26.

41. Malhotra, R., & Malhotra, D. K. (2003). Evaluating consumer loans using neural networks. *Omega*, 31(2), 83-96.

42. West, D. (2000). Neural network credit scoring models. *Computers & Operations Research*, 27(11-12), 1131-1152.

43. Eck, D., & Schmidhuber, J. (2002). A first look at music composition using lstm recurrent neural networks. *Istituto Dalle Molle Di Studi Sull Intelligenza Artificiale*, 103, 48.

44. Pomerleau, D. A. (1991). Efficient training of artificial neural networks for autonomous navigation. *Neural Computation*, 3(1), 88-97.

45. Fierro, R., & Lewis, F. L. (1998). Control of a nonholonomic mobile robot using neural networks. *IEEE Transactions on Neural Networks*, 9(4), 589-600.

46. Miyamoto, H., Kawato, M., Setoyama, T., & Suzuki, R. (1988). Feedback-error-learning neural network for trajectory control of a robotic manipulator. *Neural Networks*, 1(3), 251-265.

47. Reinelt, G. (2003). *The traveling salesman: computational solutions for TSP applications* (Vol. 840). Springer-Verlag.

48. Schalkoff, R. J. (1997). *Artificial neural networks*. McGraw-Hill Higher Education.

49. Malik, M., Nandal, R., Dalal, S., Jalglan, V., & Le, D. N. (2022). Deriving driver behavioral pattern analysis and performance using neural network approaches. *Intelligent Automation & Soft Computing*, 32(1), 87-99.

50. Zirilli, J. S. (1997). *Financial prediction using neural networks* (Vol. 254). UK: International Thomson Computer Press.

Editors

Kanta Prasad Sharma, PhD
Computer Science Engineering Department
G L Bajaj Group of Institutes, Mathura, Uttar Pradesh, India

Dr. Kanta Prasad Sharma has a rich experience of more than 13 years as an innovative academician. He has a PhD in Information Technology from Amity University & Masters from UPTU (Uttar Pradesh Technical University) Lucknow, India. At present, Dr Sharma is the Assistant Professor in Computer Science & Engineering, and coordinator of the R&D Cell at GL BAJAJ Group of Institutions. Dr. Sharma has published 15+ research papers in SCI & Scopus Indexed Conferences & Scopus Indexed Journals and also is a member of ACM, Computer Society of India, IEEE and an advisory board member of ERDA. Dr. Sharma has been engaged in guiding undergraduate students and is a research scholar co-guide in the field of real time navigation, wireless sensor networks, touch control technology & IoT, and cloud computing. He is an editorial board member of more than five international and national journals. Dr. Sharma is working as the Guest Editor of Springer Journal & Springer book editor (Scopus indexing) and is also involved as a TPC member in more than seven international conferences. He has delivered more than 10 special talks and managed more than 5 special session in international conferences. He has received the best research paper award in international conferences at Amity campus Dubai. Dr. Sharma has visited more than five countries for research and academic engagements.

Shaurya Gupta, PhD
School of Computer and System Science
University of Petroleum and Energy Studies, Dehradun, Uttarakhand, India

Dr. Shaurya Gupta is currently working in the capacity of Assistant Professor at the University of Petroleum and Energy Studies, Dehradun, Uttarakhand, India. Prior to this, he worked at Amity University, Rajasthan, India, in the same capacity. He has completed his PhD (IT) from Amity University, Rajasthan, and received a MTech (CS) from Jagannath University Jaipur and has over 11 years of experience in teaching, administration, liaison & coordination, student management and research & analysis. Dr. Gupta has published various research papers in reputed research journals and attended many national and three international conferences. His areas of interest are delay tolerant network, wireless sensor network, ad hoc network, IoT and machine learning. Apart from research, he is involved in academics by means of preparing exercises, questionnaires and assignments for students at various levels and setting and marking assignments and tests, and assessing students work for internally assessed components of qualifications at both UG and PG level.

Ashish Sharma, PhD
School of Computer Science
GLA University, Mathura (UP), Chaumuhan, Uttar Pradesh, India

Dr. Ashish Sharma received a PhD in Computer Science with a concentration in Data Mining. Apart from teaching graduate classes, Dr. Sharma is working on the applications of data mining, machine learning, big data, business analytics, data/text mining, healthcare informatics, and service-oriented computing. He is one of the key faculty members of the information retrieval and data mining group of GLA University, India. Dr. Sharma has developed a new approach to the study and development of facility location problems using data mining algorithms, where the objective is to develop a good decision algorithm for the better allocation of facility location problems. He and his team have filed a patent on "IoT Enabled Solar Power Based Real-time Environment Monitoring Device," Patent No: 201811013241A. In 2013, he won the Best Mentor Award given by IBM India Pvt. Ltd at the national level for developing various projects for the national level contest "The Great Mind Challenge" of IBM. He has also chaired various conferences organized by IEEE at GLA University. Recently, his paper on burn detection was accepted in the IEEE journal with impact factor 2.075; in addition to this, he has published various papers in Scopus indexes journals and conference. He is very keen to do further research work to develop a machine learning model that will account for various user parameters like medical imaging, MRI and X-ray images. Recently, he has completed his work on the Burn Images, Algae Identification, Bone Fracture Detection, Cardiac data set and now is working on the Bone cancer data set.

Dac-Nhuong Le, PhD
Associate Professor in Computer Science
Head of Faculty of Information Technology
Haiphong University, Haiphong, Vietnam

Dac-Nhuong Le obtained an MSc and PhD in computer science from Vietnam National University, Vietnam, in 2009 and 2015, respectively. He is an Associate Professor in Computer Science and Head of Faculty of Information Technology, Haiphong University, Vietnam. He has a total academic teaching experience of 15+ years and has been published 100+ reputed international conferences, journals and online book chapter contributions (Indexed by: SCIE, SSCI, Scopus, ACM, DBLP). His areas of research include soft computing, network communication, security and vulnerability, network performance analysis and simulation, and cloud computing. His core work is in network security, soft computing and IoT and image processing in the biomedical field. Recently, he was a member of the technique program committee that reviewed techniques. He was also the track chair for international conferences under Springer Series. Presently, he is serving on the editorial board of international journals and has authored/edited 20+ computer science books by Springer, Wiley, IET, and CRC Press.

Also of Interest

Check out these other similar books by Dac-Nhuong Le published by Scrivener Publishing

Evolving Networking Technologies
Developments and Future Directions
Edited by Kanta Prasad Sharma, Shaurya Gupta, Ashish Sharma and Dac-Nhuong Le
Published 2023. ISBN 978-1-119-83620-9

Fuzzy Logic Applications in Computer Science and Mathematics
Edited by Rahul Kar, Dac-Nhuong Le, Gunjan Mukherjee and Biswadip Basu Mallik and Ashok Kumar Shaw
Published 2023. ISBN 978-1-394-17453-9

Evolving Software Processes
Trends and Future Directions
Edited by Arif Ali Khan and Dac-Nuong Le
Published 2022 ISBN 978-1-119-82126-7

Cyber Security and Digital Forensics
Challenges and Future Trends
Edited by Mangesh M. Ghonge, Sabyasachi Pramanik, Ramchandra Mangrulkar, and Dac-Nhuong Le
Published 2022. ISBN 978-1-119-79563-6

Cloud Computing Solutions
Architecture, Data Storage, Implementation and Security
Edited by Souvik Pal, Dac-Nhuong Le and Prasant Kumar Pattnaik
Published 2022. ISBN 978-1-119-68165-6

Cryptocurrencies and Blockchain Technology
Decentralization and Smart Contracts
Edited by Gulshan Shrivastava, Dac-Nhuong Le, and Kavita Sharma
Published 2020. ISBN 978-1-119-62116-4

Ontology-Based Information Retrieval for Healthcare Systems
Edited by Vishal Jain, Ritika Wason, Jyotir Moy Chatterjee and Dac-Nhuong Le
Published 2020. ISBN 978-1-119-64048-6

Machine Learning and Cognitive Computing for Mobile Communications and Wireless
Edited by Krishna Kant Singh, Akansha Singh, Korhan Cengiz and Duc-Nhuong Le
Published 2020. ISBN 978-1-119-64036-3

Emerging Extended Reality Technologies for Industry 4.0
Early Experiences with Conception, Design, Implementation, Evaluation and Deployment
Edited by Jolanda G. Tromp, Dac-Nhuong Le and Chung Van Le
Published 2020. ISBN 978-1-119-65463-6

Security Designs for the Cloud, IoT and Social Networking
Edited by Dac-Nhuong Le, Chintan Bhatt and Mani Madhukar
Published 2019. ISBN 978-1-119-59226-6

Network Modeling, Simulation and Analysis in MATLAB
Theory and Practices
Edited by Dac-Nhuong Le, Abhishek Kumar Pandey, Sairam Tadepalli, Pramod Singh Rathore and Jyotir Moy Chatterjee
Published 2019. ISBN 978-1-119-63143-9

Cyber Security in Parallel and Distributed Computing
Concepts, Techniques, Applications and Case Studies
Edited by Dac-Nhuong Le, Raghvendra Kumar, Brojo Kishore Mishra, Manju Khari and Jyotir Moy Chatterjee
Published 2019. ISBN 978-1-119-48805-7

Emerging Technologies for Health and Medicine
Virtual Reality, Augmented Reality, Artificial Intelligence, Internet of Things, Robotics, Industry 4.0
Edited by Dac-Nhuong Le, Chung Van Le, Jolanda G. Tromp and Gia Nhu Nguyen
Published 2018. ISBN 978-119-50981-3

Cloud Computing and Virtualization
Edited by Dac-Nhuong Le, Raghvendra Kumar, Gia Nhu Nguyen and Jyotir Moy Chatterjee
Published 2018. ISBN 978-119-48790-6

Printed and bound by CPI Group (UK) Ltd, Croydon, CR0 4YY

17/09/2026

14964281-0001